AF454189

PHARMACEUTICAL MARKETING 4.0 – INDIAN CONTEXT

PHARMACEUTICAL MARKETING 4.0 – INDIAN CONTEXT

Vijay Bhangale
Director,
IES's Management College and Research Centre,
Bandra Reclamation, Bandra, Mumbai - 40050

PharmaMed Press
An imprint of BSP Books Pvt. Ltd.
4-4-309/316, Giriraj Lane,
Sultan Bazar, Hyderabad - 500 095.

Pharmaceutical Marketing 4.0 – Indian Context
by **Vijay Bhangale**

> *Disclaimer:* The authors and publishers have taken due care to provide authentic, reliable and up-to-date information related to the subject. However, neither the authors nor the publisher shall be responsible for any liability for any damage caused because of the use of this book. The respective users must check the accuracy from other sources too.

Published by

PharmaMed Press
An imprint of BSP Books Pvt. Ltd.
4-4-309/316, Giriraj Lane, Sultan Bazar, Hyderabad - 500 095.
Phone: 040-23445688; Fax: 91+40-23445611
E-mail: info@pharmamedpress.net
www.bspbooks.net/www.pharmamedpress.net

ISBN: 978-81-961468-0-1 (Hardbound)

Dedication,

I would like to dedicate this book to my parents, my wife and my daughter.

Preface

With over 17 years of experience in Pharmaceutical/Healthcare industry and about 16 years of experience in academics teaching a course on Pharmaceutical Marketing to the students of a Program on Pharmaceutical & Healthcare Management, I thought of writing this book on Pharmaceutical Marketing. Although there are few books in the market on Pharmaceutical Marketing, there is a need for a book focusing on marketing concepts, especially with respect to Pharmaceutical Industry and in Indian context.

The book is application oriented highlighting the new trends within Indian Pharmaceutical Industry, Challenges that Marketers are likely to face & the Strategies that they can adopt. It is also designed keeping in mind the Pharmaceutical & Healthcare management programs taught at various B-schools. This book serves as a reference book for pharma marketers and the students of Pharmaceutical & Healthcare Management Programs.

Highlights of the Book

- Covers marketing concepts which a pharma marketer needs to know
- It focuses on latest concepts and examples
- Provides frameworks that can be applied on-the-job
- A summary is given at the end of each chapter for quick reference
- It is backed by author's Industry experience coupled with academic experience of teaching the subject to students undergoing management program in Pharmaceutical & Healthcare Management

Book is for whom?

- Pharma Industry Marketing Practitioners
- Students undergoing a Pharmaceutical /Healthcare Management Program

Chapters Structure

Chapter 1 Healthcare Scenario in India

This chapter deals with defining Health & describing the Indian healthcare industry, its segments, size, opportunity, and challenges.

Chapter 2 Indian Pharmaceutical Industry Landscape

This chapter deals with the Indian Pharmaceutical Industry landscape & an overview of the Indian Pharmaceutical Market.

Chapter 3 Understanding Pharmaceutical Customer

This chapter deals with the understanding of Pharmaceutical Customer (Doctor), Doctor's decision-making process, factors influencing doctor's prescription.

Chapter 4 Pharmaceutical Marketing – How is it different from Consumer Marketing

This chapter deals with the understanding of how pharmaceutical marketing is different from marketing consumer goods and the consumer goods marketing strategies that can be adopted by pharmaceutical marketers.

Chapter 5 Pharma 4.0

This chapter deals with the understanding of concepts of Industry 4.0, Pharma 4.0, patient centricity, tracking patient journey, patient-centered apps, and emerging technologies and their application.

Chapter 6 Marketing Research and MIS

This chapter deals with Marketing Information System - databases used in pharmaceutical industry & Marketing Research types, process and application.

Chapter 7 Identifying Market Segments, Targeting & Positioning

This chapter deals with Market Segmentation, Targeting, and Positioning, ways in which pharmaceutical market can be segmented, how to go about selecting target market and positioning a pharmaceutical brand.

Chapter 8 Pharmaceutical Brand and Branding Strategies

This chapter deals with product vs brand, brand identity, pharmaceutical brand naming, and branding strategies.

Chapter 9 New products - growth drivers in pharma

This chapter deals with the new product concept, new product development process in pharma, pharmaceutical product life cycle, and product strategies for growth.

Chapter 10 Pricing Strategies

This chapter deals with pricing, regulations regarding drug pricing, pricing methodologies, and pricing strategies.

Chapter 11 Pharmaceutical Distribution

This chapter deals with the understanding of the pharmaceutical value chain, distribution channel structure, assessing distribution channels, physical distribution and challenges, and omnichannel strategy.

Chapter 12 Importance of retail pharmacy & emergence of e-pharmacy

This chapter deals with the pharmaceutical retail universe, the growing importance of retail, the growing influence of e-pharmacy, e-pharmacy advantages, and challenges.

Chapter 13 Building Pharma Brands

This chapter deals with an understanding of brand equity, brand building process, tracking brand health, the role and responsibilities of a brand manager, and the brand planning process.

Chapter 14 Integrated Marketing Communications

This chapter deals with the concept of integrated marketing communications, Pharmaceutical promotional mix, promotional aides used in pharmaceutical marketing, designing of creative campaigns, application of emerging technologies (AR, VR, IoT), and allocation of promotional budget.

Chapter 15 Salesforce Effectiveness – key to pharmaceutical brand success

This chapter deals with an understanding of pharmaceutical selling, changing scenarios of pharmaceutical sales force requirements, sales force effectiveness in the pharmaceutical industry, and sales automation.

Chapter 16 Relationship Marketing

This chapter deals with building relationships, engaging with healthcare professionals, strategies for building relationships, sustaining and nurturing strategies & managing Key Opinion Leaders (KOLs).

Chapter 17 Medico Marketing

This chapter deals with the concept of medico marketing, the role of the medical services department, medico marketing tools, and evolving role of medico marketing.

Chapter 18 Digital Transformation in Pharmaceutical Industry

This chapter deals with the digital ecosystem, components of digital marketing, digital strategies for engaging with healthcare professionals & patients, measuring the effectiveness of digital marketing, and the near future digital outlook.

Chapter 19 Hospital (Institutional) Marketing

This chapter deals with the significance of hospital/institutional marketing, the process of institutional marketing, and institutional marketing strategies.

Chapter 20 Rural Marketing – its significance in the Indian context

This chapter deals with the rural healthcare scenario in India, the significance of rural marketing, rural marketing initiatives by pharmaceutical companies and rural marketing strategies in the pharmaceutical industry.

Chapter 21 International Marketing - exploiting the potential of generics

This chapter deals with international marketing and the international marketing environment, the global pharmaceutical market and role of the Indian pharmaceutical industry, global pharmaceutical market classification, selection of international market, internationalization of the Indian pharmaceutical industry, how the marketing mix for international marketing is different and registration of products in international markets.

Chapter 22 OTC Marketing - a growth strategy

This chapter deals with an understanding of the concept of OTC Marketing, its significance and challenges in the Indian context, the process of OTC Marketing, and Rx to OTC switch strategies.

Chapter 23 Marketing of specialty drugs for rare diseases

This chapter deals with the understanding of specialty drugs and rare diseases & marketing of specialty drugs for rare diseases.

Chapter 24 Biopharmaceuticals

This chapter deals with the understanding of biopharmaceuticals and biosimilars, market analysis, and strategies to be successful in this market.

This book can be a part of Brand Manager's library – a book s/he can refer to whenever s/he has to clarify concepts. I would be more than delighted to see this book used at B-schools running a program on pharmaceutical and healthcare management. And would love to hear the reader's feedback and correct any imperfections in future editions.

-Author

Acknowledgements

Foremost, I offer my heartfelt thanks to all my Industry Colleagues, Academic Colleagues and Students who have shared their experiences with me.

I am ever appreciative of the assistance of so many.

To Prof. Ulhas Karkhanis, Associate Dean, Pharma Management program at IES's Management College and Research Centre, Thank you for your time, efforts and energy in helping me with the chapter on Salesforce effectiveness.

To Mr. Anuraag Dalal, Chief Digital Strategist at Charak Pharma Pvt. Ltd., Thank you for sharing your knowledge and insights with me on Digital transformation in pharma industry. Truly, no words can I express my gratitude.

To Dr. Rajiv Gatne, former Director - International Business, Johnson and Johnson Ltd & CEO - Sunshine Healthcare Ltd, Accra, Republic of Ghana, big Thank you for sharing your expertise in building up the chapter on International marketing.

To Mr. Govind Pathak, student of Pharma Management program at IES's Management College and Research Centre, thank you for your inputs on Relationship Marketing.

Last but not the least I would like to express my deep gratitude to Ms. Sunita Masiwal, Stenographer at IES's Management College and Research Centre for her assistance in creating tables and flow charts.

-Author

Contents

About the Author

Dr. Vijay Bhangale (B.Pharm., M.M.S., Ph.D.)

Dr. Vijay Bhangale is a Management Consultant, Corporate Trainer and an Academician with 17 years of Industry experience & 16 years of experience in academics.

Currently, he is Director at IES's Management College and Research Centre, Mumbai, India and has taught subjects like Pharmaceutical Marketing, OTC Marketing, Brand Management, Integrated Marketing Communications and Marketing Strategy.

He has presented research papers in conferences at New York, USA, IIMA, IIMK, IIML etc. His research papers have been published in various international and national journals.

He has conducted training programs on Pharma Brand Management & has successfully carried out numerous strategic consulting projects for Pharma/Consumer healthcare companies like 'Market Entry Strategies', 'Growth Strategies', 'Business Plan for New Division' etc. Prior to getting into academics, he has worked with companies like GSK Consumer Healthcare, Ranbaxy, Ipca, Merck etc. and successfully handled marketing of Pharmaceutical, OTC and FMCG products. He headed the profit centres handling leader brands and was responsible for setting up new strategic business units & new product launches.

CHAPTER 1

HEALTHCARE SCENARIO IN INDIA

LEARNING OUTCOMES

After reading this chapter, you should be able to:

- Explain the concept of Health
- Understand different segments of the Indian Healthcare Industry
- Describe Healthcare scenario in India
- Understand the growth drivers and opportunities in Indian Healthcare Industry
- Understand challenges in Indian Healthcare Industry

INTRODUCTION

It is said that 'Health is Wealth'. Good health is important to all of us to excel in whatever we do.

The Healthcare industry is truly putting into action the idea captured in the old saying 'adding years to life and adding life to one's years.

Let us first understand the meaning of the word 'Health'.

WHAT IS 'HEALTH'?

According to World Health Organization (WHO) 'Health' is defined as follows:

Health is a state of complete physical, mental and social well-being and not merely the absence of disease or infirmity.

INDIAN HEALTHCARE INDUSTRY

Healthcare in India comprises of following key segments:

HOSPITALS

Government hospitals including primary healthcare centers, district, and general hospitals and

Private hospitals include pursing homes, mid-tier, top-tier, and super-specialized establishments.

Pharmaceuticals: Include the manufacturing, extraction, processing, purification, and packaging of chemicals to be used as medication.

Diagnostic Services: Comprising of businesses and laboratories that offer diagnostic services

Medical Devices: Includes establishments engaged in manufacturing and maintaining medical equipment for surgical, dental, ophthalmic, laboratory, etc. use.

Medical Insurance: Comprises insurance to cover hospitalization expenses, reimbursements

Telemedicine: Comprises of online or teleconsultation and is driven by the growing adoption of technology-led solutions in healthcare

Over-The-Counter (OTC) medicines: This includes consumer healthcare products that do not need a prescription and are bought by consumers on their own.

HIGHLIGHTS OF HEALTHCARE INDUSTRY IN INDIA

The key highlights of the Indian Healthcare industry are as follows:

Indian Healthcare industry is estimated to reach US$ 193.83 billion by 2020 and US$ 372 billion (INR 27 trillion) by 2022 according to IBEF (INDIA BRAND EQUITY FOUNDATION) Healthcare Report, May 2021.

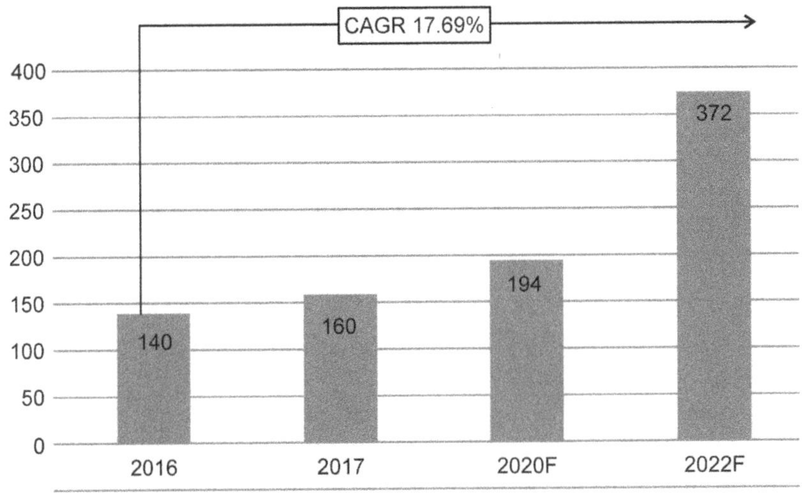

Note: F - Forecast
Values are in Billion USD

Fig. 1.1 Indian Healthcare Sector Growth Trend
Source: IBEF (INDIA BRAND EQUITY FOUNDATION) Healthcare Report,
May 2021

India is the world's second-largest populated country. From 1 billion in 2000 to 1.35 billion in 2015, the rising population demands continuous growth and improvement in healthcare facilities.

The largest contributor to the healthcare industry is the hospital sector which stood at INR 4 trillion (US$ 61.79 billion) in 2017 and is expected to increase at a Compound Annual Growth Rate (CAGR) of 16-17 percent to reach INR 8.6 trillion (US$ 132.84 billion) by 2022 (IBEF Healthcare Report, Dec 2019).

Nearly 70 percent of expenditure on health has to be borne by the patients and this pushes about 60 million Indians into poverty each year according to the Finance Commission report (The New Indian Express, Feb 28, 2021).

The percentage of government spending (including current and capital expenditure) in the health sector to GDP increased from 1.19 percent in 2015-16 to 1.33 percent in 2017-18 (Sustainable Development Goals National Indicator Framework Progress Report, 2020).

In Budget 2021, India's public expenditure on healthcare stood at 1.2% as a percentage of the GDP. The Government is planning to increase public health spending to 2.5% of the country's GDP by 2025 as shown in the chart below.

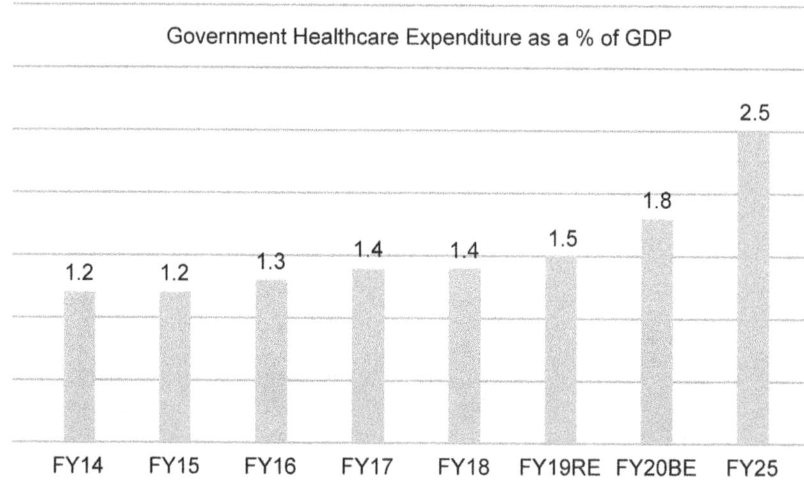

Note: BE - Budget Estimated, RE-Revised Estimate, World Bank, Economic Survey FY20

Fig. 1.2 Government Health Expenditure
Source: IBEF (INDIA BRAND EQUITY FOUNDATION) Healthcare Report, May 2021

The Pharmaceutical industry (including drugs & medical devices) is around US$ 43 billion (Rs.3, 01,000 crores) and is currently having a growth rate of 7-8% in the drug sector and 15-16% in the medical device sector. Total exports (drugs and medical devices) are to the tune of US$20 billion (Rs.1,47,420 crore) of which drugs form around 90% of the total exports (Annual Report 2019-20, Department of Pharmaceuticals, Government of India).

According to Edelweiss's report (2017), the overall size of the diagnostics industry is INR 434 billion, of which diagnostic chains account for ~15%. Of this, while pan-India players' share stands at 6%, that of standalone centers and hospitals constitutes 48% and 37%, respectively.

The medical devices market is expected to cross USD 11 billion by 2022(IBEF Healthcare Report, Dec 2019).

Growth in teleconsultation peaked after the pandemic-induced lockdown and reached a market size of $163 million in March 2021. According to a recent report by Praxis Global Alliance, a management consulting and advisory services firm, the online doctor consultation market is expected to be over $800 million by FY2024 growing at 72 percent CAGR (Business Standard, Aug 21, 2021).

The Covid-19 pandemic has put healthcare on priority boosting the OTC market in India. With increasing health consciousness, greater urbanization, and higher levels of education and awareness there is rising demand for OTC products. Indian OTC market has been valued at approx. Rs.30,000 crores and is growing at a CAGR of 9% (Nicholas Hall's DB6 Global OTC Database – India, Dec 2020 MAT).

HEALTH PARAMETERS

According to United Nations Population Fund Annual (UNFPA) Report 2019, life expectancy has gone up to 69 years and 6 percent of the population (1368.7 million) that is 82 million is above 65 years. The elderly population is likely to be 173 million by 2026 is going to significantly contribute to growing healthcare spending. The key health parameters are given in the table below.

Table 1.1 Health Parameters

Sr.No.	Parameter	2017
1	Crude Birth Rate (per 1000 population)	20.2
2	Crude Death Rate (per 1000 population)	6.3
3	Total Fertility Rate (Per woman)	2.2
4	Maternal Mortality Ratio (Per 100,000 live births)	122
5	Infant Mortality Rate (Per 1000 live births)	33
6	The expectation of life at Birth(years)	69
		(2013-17)
		Mid-year 2015

SRS, Registrar General & Census Commissioner, India

Source: HEALTH AND FAMILY WELFARE STATISTICS IN INDIA 2019-20, Ministry of Health and Family Welfare Government of India

RISE IN NON-COMMUNICABLE DISEASES

In a recent report of the Indian Council of Medical Research (ICMR), titled India: Health of the Nation's States: The India State-Level Disease Burden Initiative (2017), it is observed that the disease burden due to communicable, maternal, neonatal, and nutritional diseases, as measured using Disability-adjusted life years (DALYs), dropped from 61 percent to

33 percent between 1990 and 2016. In the same period, the disease burden from non-communicable diseases increased from 30 percent to 55 percent.

HEALTHCARE INFRASTRUCTURE

According to National Health Profile 2019, the total number of registered Allopathic Doctors (up to 2018) is 11, 54,686. The number of Dental Surgeons registered with Central/State Dental Councils of India up to 31.12.2018 was 2,54,283. The total number of registered AYUSH Doctors in India as of 01.01.2018 was 7,99,879. Given below is the latest data on the state-wise number of medical colleges and allopathic doctors.

Table 1.2 State-wise Medical Colleges (as on 23-3-2020)

State	Colleges
Andaman & Nicobar Islands	1
Andhra Pradesh	30
Arunachal Pradesh	1
Assam	6
Bihar	14
Chandigarh	1
Chattisgarh	9
Dadra and Nagar Haveli	1
Delhi	9
Goa	1
Gujarat	29
Haryana	12
Himachal Pradesh	7
Jammu & Kashmir	8
Jharkhand	6
Karnataka	60
Kerala	32
Madhya Pradesh	21
Maharashtra	55
Manipur	2
Meghalaya	1
Mizoram	1

Contd…

State	Colleges
Orissa	11
Pondicherry	8
Punjab	9
Rajasthan	22
Sikkim	1
Tamil Nadu	49
Telangana	32
Tripura	2
Uttar Pradesh	55
Uttarakhand	5
West Bengal	24
INI's	16
Total	**541**

INIs (AIIMS Delhi + AIIMS at Bhopal, Bhubaneswar, Jodhpur, Patna, Raipur, Rishikesh, Manglagiri, Nagpur, Rae Bareli, Gorakhpur, Kalyani, Deoghar, Bhatinda, Bibinagar and PGIMER Chandigarh. JIPMER Puducherry)

Source: MoHFW (ME-I) / Medical Council of India, HEALTH AND FAMILY WELFARE STATISTICS IN INDIA 2019-20

Table 1.3 State-wise Number of Allopathic Doctors

Name of State Medical Council	Total Number of Allopathic Doctors
Andhra Pradesh Medical Council	104886
Arunachal Pradesh Medical Council	1246
Assam Medical Council	24083
Bihar Medical Council	45795
Chattisgarh Medical Council	10020
Delhi Medical Council	26685
Goa Medical Council	4035
Gujarat Medical Council	71348
Haryana Dental & Medical Councils	15679
Himachal Pradesh Medical Council	3406
Jammu & Kashmir	16648
Jharkhand Medical Council	6926
Karnataka Medical Council	131903
Madhya Pradesh Medical Council	42600
Maharashtra Medical Council	188540

Contd…

Name of State Medical Council	Total Number of Allopathic Doctors
Medical Council of India	52667
Mizoram Medical Council	118
Nagaland Medical Council	141
Orissa Council of Medical Registration	24780
Punjab Medical Council	51685
Rajasthan Medical Council	48230
Sikkim Medical Council	1414
Tamil Nadu Medical Council	148216
Travancore Medical Council, Cochin	65672
Uttar Pradesh Medical Council	84560
Uttaranchal Medical Council	9348
West Bengal Medical Council	77664
Tripura Medical Council	1945
Telengana Medical Council	7932
Grand Total	**1268172**

Source: GOVERNMENT OF INDIA MINISTRY OF HEALTH AND FAMILY WELFARE DEPARTMENT OF HEALTH AND FAMILY WELFARE, July 2021

This data is extremely important in pharmaceutical marketing as it gives an idea of the universe of customers (doctors), which will be explained in later chapters of this book.

HEALTH INSURANCE

Health insurance in India is a growing segment. The advent of private insurers in India saw the introduction of many innovative products like family floater plans, top-up plans, critical illness plans, hospital cash, and top-up policies. Ayushman Bharat Mission- National Health Protection Mission or Pradhan Mantri Jan Arogya Yojana (PMJAY) world's largest health scheme announced in the Union Budget 2018-19 is the latest initiative in expanding the health insurance net and targets 10 crores poor and deprived rural population. The Mission aims to provide a cover of Rs.5 lakh per family per year for secondary and tertiary care procedures.

CGHS - Categories of Beneficiaries

Central Government Health Scheme (CGHS) Central Government Health Scheme (CGHS) is a health scheme for serving/retired Central Government employees and their families.

Presently approximately 38.5 lakh beneficiaries are covered by CGHS in 74 cities all over India and the endeavor is to include more cities to improve the accessibility of the services (*https://cghs.gov.in).*

Employees' State Insurance Scheme (ESIS)

The Employees' State Insurance Scheme is an integrated measure of Social Insurance embodied in the Employees' State Insurance Act and it is designed to accomplish the task of protecting 'employees' as defined in the Employees' State Insurance Act, 1948 against the impact of incidences of sickness, maternity, disablement and death due to employment injury and to provide medical care to insured persons and their families. The Act now applies to over 7.83 lakhs factories and establishments across the country, benefiting about 2.13 crores insured persons/ family units. As of now, the total beneficiary stands at over 8.28 crores (*https://www.india.gov.in).*

Growth drivers and opportunities in Indian Healthcare Industry

1. **Demographic Trends:** Demographic trends like rising income, growing middle class, increase in life expectancy, a growing number of senior citizens (60+ years) are likely to boost the demand for healthcare as well as influence the nature of health services demanded in the years to come.

2. **Epidemiological Trends:** Certain epidemiological trends like a dual burden of communicable as well as non-communicable diseases (NCDs) are presenting a huge opportunity for the healthcare industry.

   India currently has around 60 million diabetics, a number that is expected to swell to 90 million by 2025(IBEF Healthcare report, July 2019). It is estimated that every fourth individual in India aged above 18 years has hypertension (KPMG-FICCI report on Healthcare, Jan10,2021 & PwC report on funding Indian Healthcare, Dec16,2020). Nearly 5.8 million Indians die from NCDs (heart and lung diseases, stroke, cancer, and diabetes) every year (PwC report on Financing and Funding Indian Healthcare, Dec16,2020 & IBEF Healthcare report, July 2019). The rising NCD burden is estimated to cost India USD 4.58 Trillion before 2030 (IBEF Healthcare report, July 2019).

3. **Demand-supply gap:** India currently has 8.6 doctors and 5 hospital beds respectively per 10,000 population (The Times of India,

Dec17,2020). While 70 percent of the population lives in rural areas, 80 percent of doctors work in cities (Dimple Vij, Int J Multidisciplinary, Mar 2019) and 69 percent of hospital beds are concentrated in urban areas (The Economic Times, May 12, 2021). An additional 3 million beds will be needed for India to achieve the target of 3 beds per 1,000 people by 2025 and another 1.54 million doctors and 2.4 million nurses will be required to meet the growing demand for healthcare in India as per NITI Aayog's report on Investment Opportunities in India's Healthcare Sector, March 2021.

4. **Medical Tourism:** India is fast emerging as an attractive destination for medical value travelers from across the globe. In particular, wellness tourism is growing faster than global tourism, as an increasing number of consumers are incorporating wellness into their travel plans.

5. **Telemedicine and adoption of digital technology:** According to a study (survey of consumers, doctors, and stakeholders from pharma companies and global EY research) by EY, in collaboration with the Indian Pharmaceutical Alliance (IPA) in September 2020, the domestic telemedicine market is expected to reach US$ 5.5 billion by 2025. By April 2021, the Health Ministry's eSanjeevani telemedicine service crossed 5 million (50 lakh) teleconsultations since its launch, enabling patient-to-doctor consultations, from the confines of their home, and doctor-to-doctor consultations (IBEF Healthcare Report, May 2021).

6. **Government initiatives:** Several policy measures taken by the government are likely to drive the growth of India's healthcare sector. These include an increase in public health expenditure to 2.5% of GDP by 2025; implementation of several large-scale and ambitious initiatives like Ayushman Bharat; commitment from the Government to invest USD 200 Billion in medical infrastructure by 2024 as well as the roll-out of various schemes under the AatmaNirbhar Bharat Abhiyaan. The Performance-Linked Incentive (PLI) Scheme and the Scheme for Promotion of Medical Device Parks, in particular, offer significant financial incentives for investors to manufacture in India. A new PLI 2.0 scheme is also being prepared for the promotion of the in-vitro diagnostics market (NITI Aayog's report on Investment Opportunities in India's Healthcare Sector, March 2021).

HEALTHCARE CHALLENGES

Despite the tremendous growth opportunity, there are several challenges. India ranks 154 among 195 countries in the healthcare index as per a recent survey by the Global Burden of Disease Survey (Global Burden of Disease Survey, 2016). A staggering 70 percent of the Indian population (around 700 million) lives in rural areas with restricted access to medical care. The lack of quality infrastructure, limited access to qualified medical functionaries, basic medication, and medical facilities limit the quality of health care that this population receives. Although there are many beneficial programs, they have not yielded the expected results due to the lack of systematic processes for proper data management and gaps in the implementation.

Some other challenges as stated in the PwC report (2018) are as follows:

The Indian healthcare system is reactive

Patients don't act proactively and often visit a hospital only when the disease has reached an advanced stage. This can be attributed to a lack of awareness about diseases, care, and services available.

Absence of a healthcare regulatory body

There is no healthcare regulatory body in India. In the absence of adequate regulations, there is concern among healthcare providers regarding the reliability of medical devices and thus practitioners prefer to use traditional medical devices. Although the Medical Devices Rules, 2017, became effective on 1 January 2018, their effectiveness is yet to be seen.

SUMMARY

This chapter deals with defining Health & describing the Indian healthcare industry, its segments, size, opportunity, and challenges. Key points discussed in this chapter are as follows:

1. Indian Healthcare industry comprises of key segments like Hospitals, Pharmaceuticals, Diagnostic Services, Medical Devices, Medical Insurance, Telemedicine & OTC medicines.

2. Indian Healthcare industry is likely to grow from USD 160 billion in 2017 at a CAGR of 16 percent to reach USD 372 billion (INR 26 trillion) by 2022.

3. The hospital industry in India stood at INR 4 trillion (US$ 61.79 billion) in 2017 and is expected to increase at a Compound

Annual Growth Rate (CAGR) of 16-17 percent to reach INR 8.6 trillion (US$ 132.84 billion) by 2022.

4. The Pharmaceutical industry (including drugs & medical devices) is around US$43 billion (Rs.3, 01,000 crores) and is currently having a growth rate of 7-8% in the drug sector and 15-16% in the medical device sector. Total exports (drugs and medical devices) are to the tune of US$20 billion (Rs.1, 47,420 crores) of which drugs form around 90% of the total exports.

5. The overall size of the diagnostics industry is INR 434 billion, of which diagnostic chains account for ~15%. Of this, while pan-India players' share stands at 6%, that of standalone centers and hospitals constitutes 48% and 37%, respectively.

6. The medical devices market is expected to cross USD 11 billion by 2022.

7. The online doctor consultation market is expected to be over $800 million by FY2024 growing at 72 percent CAGR.

8. Indian OTC market has been valued at approx. Rs.30,000 crores and is growing at a CAGR of 9%.

9. Life expectancy is around 69 years and the elderly population is likely to be 173 million by 2026 which is going to significantly contribute to growing healthcare spending.

10. The percentage of government spending in the health sector to GDP increased from 1.19 percent in 2015-16 to 1.33 percent in 2017-18. Out-of-pocket expenditure (OOPE) constitutes more than 60% of all health expenses.

11. The disease burden from non-communicable diseases has increased from 30 percent to 55 percent during the period 1990 to 2016.

12. The number of registered Allopathic Doctors is 12,68,172, the number of Dental Surgeons is 2, 54,283 and the number of AYUSH Doctors is 7, 99,879.

13. Government schemes on healthcare expenditure include CGHS, ESIS, Ayushman Bharat Mission, etc.

14. The healthcare challenges are the lack of quality infrastructure, limited access to qualified medical functionaries, basic medication, and medical facilities, especially in rural areas.

REFERENCES

1. www.who.int

2. IBEF (INDIA BRAND EQUITY FOUNDATION) Healthcare Report, May 2021.

3. IBEF (INDIA BRAND EQUITY FOUNDATION) Healthcare Report, Dec 2019.

4. Kumar Vikram, Nearly 70 percent expenditure on health to come out of patients' pockets: Finance Commission report, The New Indian Express, Feb 28, 2021

5. Annual Report (2019-20), Government of India, Ministry of Chemicals & Fertilizers, Department of Pharmaceuticals.

6. Edelweiss 2017 Report, Analysis Beyond Consensus, The Shift – unorganized to organized.

7. Abrar Peerzada, Doctor consultations exceed 4 billion in India in FY21: Praxis report, Business Standard, Aug 21, 2021.

8. Nicholas Hall's DB6 Global OTC Database – India, Dec 2020 MAT

9. United Nations Population Fund Annual (UNPFA) Report 2019

10. MoHFW (ME-I) / Medical Council of India, HEALTH AND FAMILY WELFARE STATISTICS IN INDIA 2019-20.

11. Indian Council of Medical Research (ICMR) Report titled India: Health of the Nation's States: The India State-Level Disease Burden Initiative (2017).

12. Government of India National Health Profile 2019

13. GOVERNMENT OF INDIA MINISTRY OF HEALTH AND FAMILY WELFARE DEPARTMENT OF HEALTH AND FAMILY WELFARE, July 2021.

14. https://cghs.gov.in

15. https://www.india.gov.in

16. IBEF (INDIA BRAND EQUITY FOUNDATION) Healthcare Report, July 2019.

17. KPMG-FICCI report on Healthcare, Jan10,2021

18. PwC report on funding Indian Healthcare, Dec16,2020

19. Nagarajan Rema, 5 hospital beds/10k population: India ranks 155th in 167, The Times of India, Dec 17, 2020.

20. Vij D. Health care infrastructure in India: Need for reallocation and regulation. International Journal of Multidisciplinary, Vol.4, Issue 3, March 2019.

21. Haidar Faizan, 69 percent of hospital beds in India are concentrated in urban areas: Report, The Economic Times, May 12, 2021.

22. NITI Aayog's report on Investment Opportunities in India's Healthcare Sector, March 2021.

23. PwC Report 2018, Reimagining the possible in the Indian healthcare ecosystem with emerging technologies.

C H A P T E R **2**

INDIAN PHARMACEUTICAL INDUSTRY LANDSCAPE

LEARNING OUTCOMES

After reading this chapter, you should be able to:

- Get an insight into Indian Pharmaceutical Industry
- Get an overview of the Indian Pharmaceutical Market
- Understand the Policy & Regulatory Landscape
- Gain knowledge about the Strengths & Weaknesses of the Indian Pharmaceutical Industry
- Understand the Opportunities & Threats for Indian Pharmaceutical Industry
- Understand the impact of the Covid-19 pandemic on the pharma industry

INTRODUCTION

Indian pharmaceutical industry is the third-largest in the World in terms of volume and tenth largest in terms of value according to the annual report 2019-20 of Government of India, Ministry of Chemicals & Fertilizers, Department of Pharmaceuticals. The total size of the industry (including drugs & medical devices) is around INR 3, 01,000 crores (US$43 billion) and is growing at the rate of 7-8% in the drug sector and 15-16% in the medical device sector. Total exports (drugs and medical devices) are to the tune of INR 1, 47,420 crores (US$20 billion) of which drugs form around 90% of the total exports (Annual report 2019-20 of Government of India, Ministry of Chemicals & Fertilizers, Department of Pharmaceuticals).

According to EY-FICCI Indian Pharma Report, Feb 2021, the Indian pharma industry is about 41.7 billion USD and comprises a domestic market worth 21 billion USD and exports worth 20.7 billion USD. The domestic market is dominated by Indian companies with an 80 percent (16.9 billion USD) share whereas global multinational companies have a share of 20 percent (4.1 billion USD). In terms of exports, the US is the largest contributor with 30.4% followed by the UK, South Africa, Russia, and Brazil as is evident from the figure below.

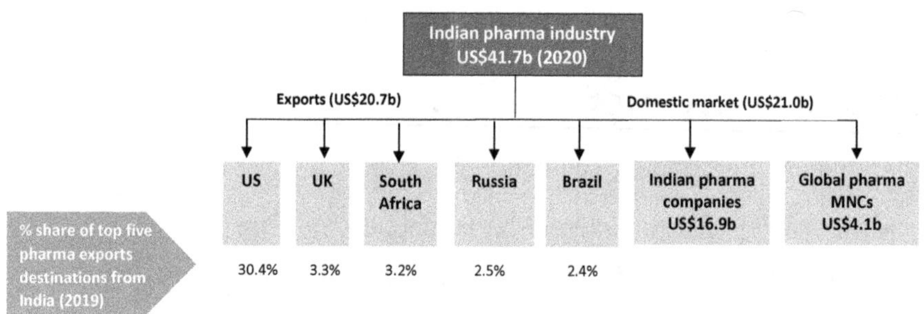

Fig. 2.1 Indian Pharma Industry

Source: EY – FICCI Indian Pharma Report, Feb 2021

India is the largest provider of generic drugs globally. Access to affordable HIV treatment from

India is one of the greatest success stories in medicine. India is one of the biggest suppliers of low-cost vaccines in the world. Because of low price and high quality, Indian medicines are preferred worldwide, thereby rightly naming the country *"the pharmacy of the world"*. The Pharmaceutical sector currently contributes around 1.72% to the country's GDP.

The Indian pharmaceutical industry has a strong network of 3000 drug companies and about 10,500 manufacturing units. Out of these, more than 2,000 units are World Health Organization (WHO) good manufacturing practice (GMP) approved; 253 are European Directorate of Quality Medicines (EDQM)-approved plants; 1,105 have Europe's Certificate of Suitability (CEPs); more than 950 match therapeutic goods administration

(TGA) guidelines; and 584 sites are approved by the US Food and Drug Administration (US FDA) (India Briefing, April 13, 2021).

INDIAN PHARMACEUTICAL INDUSTRY & GLOBAL HEALTH

Indian Pharmaceutical Industry has significantly contributed to global health outcomes. Estimates suggest that the industry, directly and indirectly, employs over 2.7 million people, in high-skill areas like R&D and manufacturing (Indian life sciences: Vision 2030, FICCI June 2015). The industry generates over USD 11 billion of trade surplus every year and is amongst the top five sectors contributing to the reduction of India's trade deficit (PHARMEXCIL, IDMA report on "Journey towards Pharma 2020 & beyond").

The Indian pharmaceutical industry is one of the top 10 sectors attracting Foreign Direct Investment (FDI) and the FDI inflow in FY 2020-21 was INR 11,015 crores (Fact sheet on FDI from April 2000 to June 2021, Department of Industrial Policy & Promotion).

Globally, Indian pharma has contributed to improved public health outcomes. India accounts for 60 percent of global vaccine production, contributing 40 to 70 percent of the WHO demand Diphtheria, Tetanus and Pertussis (DPT) and Bacillus Calmette–Guérin (BCG) vaccines, and 90 percent of the WHO demand the measles vaccine ("Affordable Efficacious Medicines – All Roads Leads to India", 2013 report by IDMA). Estimates suggest that one in every three pills consumed in the United States is produced by an Indian generic manufacturer (IQVIA 2019 report). In the UK, approximately 25 percent of the medicines used are made in India (India Brand Equity Foundation, Indian Pharmaceuticals Industry Report, Aug 2021). In Africa, the availability of affordable Indian drugs contributed to greater access to treatment for AIDS, with 37 percent of AIDS patients receiving treatment in 2009 compared to just two percent in 2003(African Businesses Magazine, 19 January 2012).

Indian Pharmaceutical Market Overview

Indian Pharmaceutical Market is valued at INR 145,354 crores according to AIOCD AWACS MAT (Moving Annual Total) Dec 2020 data and has registered the growth of 3.1 % over the same period the previous year. The decline in growth rate is due to the Covid-19 pandemic.

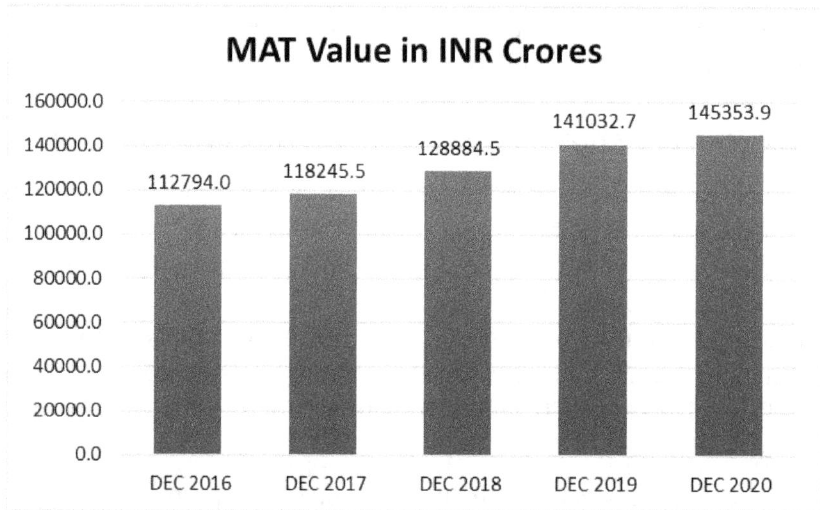

Fig. 2.2 Indian Pharma Market Revenue Trend
Source: AIOCD AWACS Pharma Industry Report, MAT Dec 2020

Indian Pharmaceutical Market is one of the fastest-growing markets globally. The market has been consistently growing as evident from the revenue trend for five years shown in the chart above.

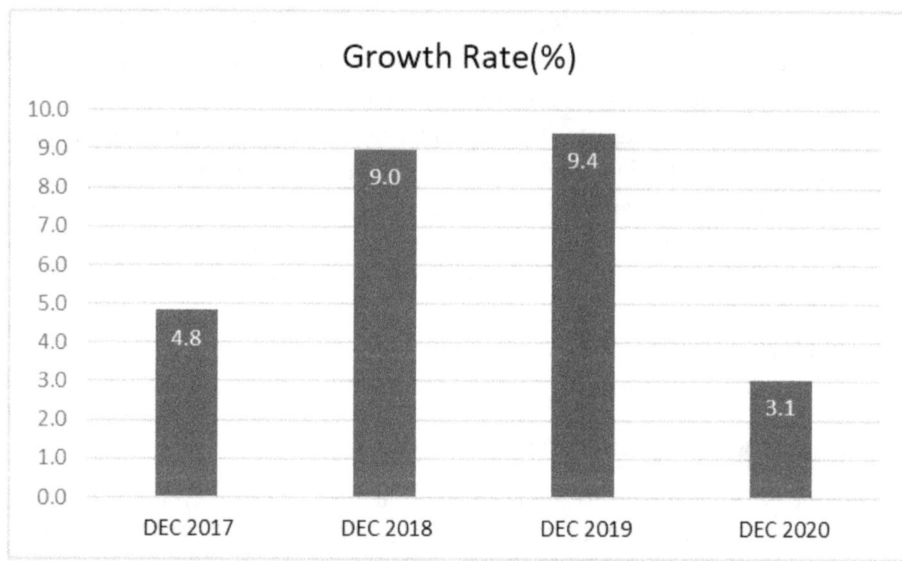

Fig. 2.3 Indian Pharma Market Growth Trend

Source: AIOCD AWACS Pharma Industry Report, MAT Dec 2020

The market growth had dipped in 2017 due to demonetization and GST roll-out but picked up in 2018 and 2019. In the year 2020, the growth rate again dipped significantly due to the Covid-19 pandemic.

The table below gives the top 50 pharmaceutical companies by sales revenue along with their market share and growth percentage.

Table 2.1 Top 50 Pharma Companies

COMPANY	RANK	MAT Value in INR Crores			Market Share
		DEC 2019	DEC 2020	Growth (%)	(%)
IPM	**0**	**141032.7**	**145353.9**	**3.1**	**100**
SUN PHARMA LABORATORIES LTD.	1	7277.3	7702.3	5.8	5.30
CIPLA LTD.	2	6624.6	7167.8	8.2	4.93
MANKIND PHARMACEUTICAL S LTD.	3	5843.8	6192.2	6.0	4.26
LUPIN LTD	4	5340.1	5551.9	4.0	3.82
ZYDUS CADILA - ZHL	5	4842.4	5068.7	4.7	3.49
ALKEM LABORATORIES LTD.	6	4529.0	4572.3	1.0	3.15
TORRENT PHARMACEUTICAL S LTD.	7	4253.7	4546.0	6.9	3.13
INTAS PHARMACEUTICAL S LTD	8	4183.4	4369.4	4.4	3.01
ABBOTT HEALTHCARE PVT. LTD	9	4209.5	4319.9	2.6	2.97
DR. REDDYS LABORATORIES LTD	10	4129.0	4216.9	2.1	2.90
MACLEODS PHARMACEUTICAL S PVT.LTD	11	3969.8	4180.2	5.3	2.88

Contd…

COMPANY	RANK	MAT Value in INR Crores			Market Share
		DEC 2019	DEC 2020	Growth (%)	(%)
IPM	0	141032.7	145353.9	3.1	100
RANBAXY LABORATORIES LTD	12	4066.3	4144.2	1.9	2.85
ARISTO PHARMACEUTICAL S PVT.LTD	13	3586.7	3908.9	9.0	2.69
GLAXOSMITHKLINE PHARMACEUTICAL S LTD.	14	4039.5	3820.1	-5.4	2.63
GLENMARK PHARMACEUTICAL S LTD.	15	3133.8	3639.4	16.1	2.50
PFIZER LTD	16	3233.8	3473.0	7.4	2.39
ABBOTT INDIA LTD.	17	3165.8	3328.1	5.1	2.29
SANOFI INDIA LTD.	18	3185.2	3313.1	4.0	2.28
USV PVT LTD	19	2739.7	3015.2	10.1	2.07
MICRO LABS LTD	20	2427.8	2534.8	4.4	1.74
IPCA LABORATORIES PVT LTD.	21	2024.6	2255.3	11.4	1.55
EMCURE PHARMACEUTICAL S LTD	22	2098.8	2208.7	5.2	1.52
ALEMBIC LTD	23	1684.3	1697.2	0.8	1.17
ZUVENTUS HEALTHCARE LTD	24	1559.2	1687.7	8.2	1.16
NOVO NORDISK INDIA PVT LTD	25	1462.7	1556.1	6.4	1.07
ERIS LIFESCIENCES LTD	26	1314.4	1435.8	9.2	0.99
HIMALAYA DRUG COMPANY	27	1211.8	1374.9	13.5	0.95
CADILA PHARMACEUTICAL S LTD	28	1043.2	1144.7	9.7	0.79
FDC LTD.	29	1123.4	1129.9	0.6	0.78

Contd…

COMPANY	RANK	MAT Value in INR Crores			Market Share
		DEC 2019	DEC 2020	Growth (%)	(%)
IPM	**0**	**141032.7**	**145353.9**	**3.1**	**100**
ZYDUS CADILA - BIOLOGICS	30	1043.4	1053.8	1.0	0.73
NOVARTIS INDIA LTD	31	1146.1	1027.8	-10.3	0.71
FRANCO INDIAN PHARMACEUTICAL S PVT LTD	32	890.5	998.0	12.1	0.69
NATCO PHARMA LTD	33	1052.1	945.6	-10.1	0.65
AJANTA PHARMA LTD	34	850.8	936.0	10.0	0.64
INDOCO REMEDIES LTD	35	940.0	905.4	-3.7	0.62
JB CHEMICALS	36	729.1	833.2	14.3	0.57
MSD PHARMACEUTICAL S PRIVATE LTD.	37	834.8	813.2	-2.6	0.56
HETERO HEALTHCARE LTD	38	858.6	795.2	-7.4	0.55
BOEHRINGER INGELHEIM	39	687.7	787.1	14.4	0.54
LA RENON	40	790.2	772.4	-2.2	0.53
MEDLEY PHARMACE-UTICALS	41	679.3	771.4	13.6	0.53
BLUE CROSS LABORATORIES LTD	42	684.7	763.4	11.5	0.53
BHARAT SERUMS & VACCINES LTD	43	695.2	730.4	5.1	0.50
HEGDE & HEGDE	44	710.7	721.6	1.5	0.50
PROCTER & GAMBLE HEALTH LTD	45	689.2	684.1	-0.7	0.47

Contd…

COMPANY	RANK	MAT Value in INR Crores			Market Share
		DEC 2019	DEC 2020	Growth (%)	(%)
IPM	**0**	**141032.7**	**145353.9**	**3.1**	**100**
APEX LABORATORIES LTD.	46	570.7	674.9	18.3	0.46
SYSTOPIC LABORATORIES LTD-MEM	47	656.0	670.5	2.2	0.46
ASTRAZENECA PHARMA INDIA LTD	48	641.3	661.6	3.2	0.46
CORONA	49	602.2	652.5	8.3	0.45
MEYER ORGANICS PVT. LTD	50	688.6	644.8	-6.4	0.44

Source: AIOCD AWACS Pharma Industry Report, MAT Dec 2020

The top 20 companies contribute about 61.3 % to the total pharmaceutical market whereas the top 50 companies' contribution is about 82.8%.

Glenmark, USV, Aristo, Ipca, and Cipla are among the fast-growing companies in the Top 20.

Other fast-growing companies in the Top 50 are Apex, Boehringer Ingelheim, Himalaya, Eris Lifesciences, Cadila, Franco Indian, Ajanta, JB Chemicals & Medley.

The top 10 Therapeutic segments in Indian Pharmaceutical Industry, their growth rates, and their contribution to the total market are given in the table below.

Table 2.2 Top 20 Therapeutic Categories

SUPER GROUP	MAT Value in INR Crores		
	DEC 2019	DEC 2020	Growth (%)
IPM	**141032.7**	**145353.9**	**3.1**
CARDIAC	17704.7	20079.7	13.4
ANTI-INFECTIVES	19244.9	18843.8	-2.1
GASTRO INTESTINAL	15683.1	16108.6	2.7
ANTI-DIABETIC	13703.3	14776.1	7.8

Contd...

SUPER GROUP	MAT Value in INR Crores		
	DEC 2019	**DEC 2020**	**Growth (%)**
IPM	**141032.7**	**145353.9**	**3.1**
VITAMINS / MINERALS / NUTRIENTS	12087.1	12884.7	6.6
RESPIRATORY	10575.2	10596.0	0.2
DERMA	9614.4	9684.5	0.7
PAIN / ANALGESICS	9630.3	9357.5	-2.8
NEURO / CNS	8475.9	8929.4	5.4
GYNAECOLOGICAL	6973.4	6730.9	-3.5
ANTI-NEOPLASTICS	2976.0	2880.8	-3.2
HORMONES	2582.3	2663.5	3.1
OPHTHAL / OTOLOGICALS	2560.3	2354.8	-8.0
VACCINES	2253.1	2289.7	1.6
UROLOGY	1924.7	2075.6	7.8
BLOOD-RELATED	1649.4	1636.6	-0.8
OTHERS	1294.0	1353.6	4.6
STOMATOLOGICALS	751.9	837.7	11.4
SEX STIMULANTS / REJUVENATORS	789.1	741.5	-6.0
ANTI MALARIALS	559.7	528.8	-5.5

Source: AIOCD AWACS Pharma Industry Report, MAT Dec 2020

The top 10 Therapeutic Categories contribute about 88% to the total pharmaceutical market.

Cardiac, Anti-infective, Gastrointestinal, Anti-Diabetic & Vitamins/Minerals/Nutrients are the Top 5 Therapeutic Categories and they contribute about 57% to the total pharmaceutical market.

Cardiac & Anti-Diabetic are the fast-growing Therapeutic Categories.

The top 20 Pharmaceutical Brands by their sales value, their subgroup, the company marketing these brands, and their sales value are given in the table below.

Table 2.3 Top 20 Pharma Brands

BRAND	COMPANY	RANK	MAT Value in INR Crores		
			DEC 2019	DEC 2020	Growth (%)
IPM		**0**	**141032.7**	**145353.9**	**3.1**
MIXTARD	NOVO NORDISK INDIA PVT LTD	1	546.2	550.7	0.8
GLYCOMET GP	USV PVT LTD	2	489.4	547.7	11.9
LANTUS	SANOFI INDIA LTD.	3	487.3	542.9	11.4
JANUMET	MSD PHARMACEUTICALS PRIVATE LTD.	4	477.8	468.6	-1.9
AUGMENTIN	GLAXOSMITHKLINE PHARMACEUTICALS LTD.	5	443.4	445.2	0.4
LIV.52	HIMALAYA DRUG COMPANY	6	405.0	441.8	9.1
BETADINE	WIN-MEDICARE PVT. LTD.	7	351.5	395.3	12.4
BECOSULES	PFIZER LTD	8	318.1	383.6	20.6
ZINCOVIT	APEX LABORATORIES LTD.	9	226.8	379.9	67.5
CLAVAM	ALKEM LABORATORIES LTD.	10	379.0	362.2	-4.4
MONOCEF	ARISTO PHARMACEUTICALS PVT.LTD	11	363.5	360.4	-0.9
THYRONORM	ABBOTT INDIA LTD.	12	351.9	352.7	0.2
FABIFLU	GLENMARK PHARMACEUTICALS LTD.	13	0.0	338.0	0.0
TELMA	GLENMARK PHARMACEUTICALS LTD.	14	272.3	336.1	23.5
FORACORT	CIPLA LTD.	15	325.0	334.5	2.9
ACILOC	CADILA PHARMACEUTICALS LTD	16	257.0	330.8	28.7
PAN	ALKEM LABORATORIES LTD.	17	300.5	328.9	9.4

Contd...

BRAND	COMPANY	RANK	MAT Value in INR Crores		
			DEC 2019	DEC 2020	Growth (%)
IPM		**0**	**141032.7**	**145353.9**	**3.1**
NOVOMIX	NOVO NORDISK INDIA PVT LTD	18	303.9	313.5	3.1
DEXORANGE	FRANCO INDIAN PHARMACEUTICALS PVT LTD	19	275.9	310.8	12.7
ROSUVAS	RANBAXY LABORATORIES LTD	20	274.6	310.3	13.0

Source: AIOCD AWACS Pharma Industry Report, MAT Dec 2020

Mixtard is the biggest brand with a MAT value of INR 550.7 crores followed by Glycomet GP (547.7 crores), Lantus (542.9 crores), and Janumet (468.6 crores). All top 4 brands are from the Anti-Diabetic category.

Zincovit is the fastest-growing brand in the Top 20 brands, primarily due to the Covid-19 pandemic as zinc supplements were recommended for enhancing immunity.

POLICY & REGULATORY LANDSCAPE

The Central Drugs Standard Control Organization (CDSCO) under the Directorate General of Health Services, Ministry of Health & Family Welfare, Government of India is the National Regulatory Authority (NRA) of India. The Drugs & Cosmetics Act, 1940 and rules 1945 have entrusted various responsibilities to central & state regulators for regulation of drugs & cosmetics ensuring the safety, rights, and well-being of the patients by regulating the drugs and cosmetics.

Important pharma regulations in India are listed in the table below.

Table 2.4 Important Pharma regulations in India

CDSCO	Central Drugs Standard Control Organization (CDSCO), Ministry of Health & Family Welfare, Government of India provide general information about drug regulatory requirements in India.
NPPA	**Drugs (Price Control) Order 1995** and other orders enforced by **National Pharmaceutical Pricing Authority (NPPA),** Government of India.

Contd...

Drugs & Cosmetics Act	**The Drugs & Cosmetics Act, 1940 regulates the import, manufacture, distribution, and sale of drugs in India.**
Schedule M	**Schedule M** of the D&C Act specifies the general and specific requirements for factory premises and materials, plant and equipment, and minimum recommended areas for basic installation for certain categories of drugs.
Schedule T	**Schedule T of the D&C Act prescribes GMP specifications for the manufacture of Ayurvedic, Siddha, and Unani medicines.**
Schedule Y	The clinical trials legislative requirements are guided by specifications of **Schedule Y** of the D&C Act.
Good Clinical Practice (GCP)	The Ministry of Health, along with the Drugs Controller General of India (DCGI) and the Indian Council for Medical Research (ICMR) has come out with draft guidelines for research in human subjects. These GCP guidelines are essentially based on the Declaration of Helsinki, WHO guidelines, and ICH requirements for good clinical practice.
Pharmacy Act	**The Pharmacy Act, 1948** is meant to regulate the profession of Pharmacy in India.
Drugs and Magic Remedies (Objectionable Advertisement) Act	**The Drugs and Magic Remedies (Objectionable Advertisement) Act, 1954 provides to control the advertisements regarding drugs; it prohibits the advertising of remedies alleged to possess magic qualities.**
The Narcotic Drugs and Psychotropic Substances Act	**The Narcotic Drugs and Psychotropic Substances Act, 1985** is an act concerned with control and regulation of operations relating to Narcotic Drugs and Psychotropic Substances.

Source: Adapted from The Indian Pharmaceutical Association (IPA) website

DRUG APPROVAL PROCESS IN INDIA

(Adapted from Sawant et al., Pharmaceut Reg Affairs, 2018 & www.cdsco.gov.in)

Under the Drugs and Cosmetics Act, CDSCO is responsible for approval of Drugs, Conduct of Clinical Trials, laying down the standards for Drugs, control over the quality of imported drugs in the country, and coordination of the activities of State Drug Control Organizations by providing expert advice with a view of bringing about the uniformity in the enforcement of the Drugs and Cosmetics Act. Further CDSCO along with state regulators, is jointly responsible for the grant of licenses of certain specialized categories of critical Drugs such as blood and blood products, I. V. Fluids, Vaccine, and Sera.

When a company in India wants to manufacture/ import a new drug it has to apply to seek permission from the licensing authority (DCGI) by filling in Form 44 also submitting the data as given in Schedule Y of Drugs and Cosmetics Act 1940 and Rules 1945. To prove its efficacy and safety in the Indian population it has to conduct clinical trials following the guidelines specified in Schedule Y and submit the report of such clinical trials in the specified format.

The regulations under the Drugs and Cosmetics Act 1940 and its rules 1945, 122A, 122B, and 122D describe the information required for approval of an application to import or manufacture a new drug for marketing.

The need for local clinical trials in India depends on the status of the drug in other countries. If the drug is already approved in other countries, generally Phase III trials are required. Phase I trials are not allowed in India unless the data is available from other countries. Permission is granted by DCGI to conduct Phase 1 trials in India, if the drug has special relevance to a health problem in India, like malaria or tuberculosis.

Bioavailability and bioequivalence (BABE) studies should be conducted as per BABE guidelines. Comprehensive information on the marketing status of the drug in other countries is also required other than the information on safety and efficacy. The information regarding the prescription, samples and testing protocols, product monographs, labels must also be submitted. It usually takes 3 months for clinical trial approval in India. The clinical trials can be registered in the Clinical Trials Registry of India (CTRI) giving details of the clinical trials and the subjects involved in the trials.

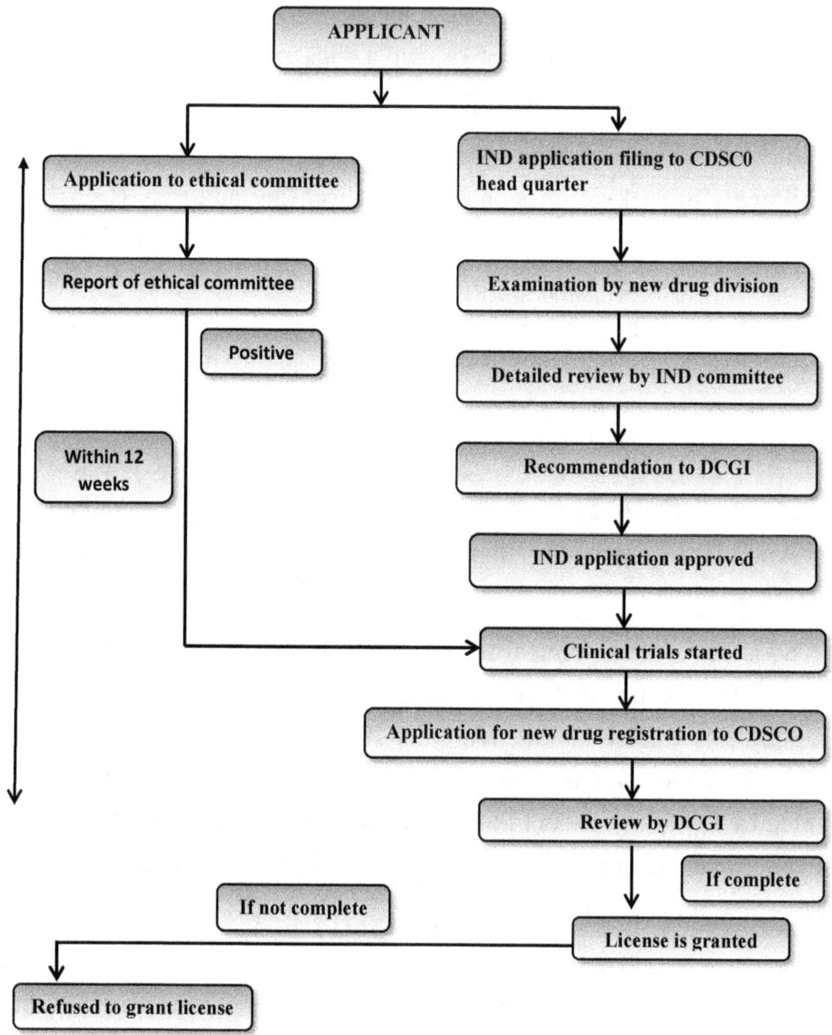

Fig. 2.4 Drug Process Approval in India
Source: Adapted from Sawant et al., Pharmaceut Reg Affairs, 2018,7:210

PRICING POLICY

(Adapted from Annual Report (2019-20), Government of India, Ministry of Chemicals & Fertilizers, Department of Pharmaceuticals).

NATIONAL PHARMACEUTICAL PRICING POLICY 2012

The Department of Pharmaceuticals notified the National Pharmaceutical Pricing Policy-2012

(NPPP-2012) on 07.12.2012 to put in place a regulatory framework for pricing of drugs to ensure availability of required medicines – "essential medicines" – at reasonable prices, while providing sufficient opportunity for innovation and competition to support the growth of industry, thereby meeting the goals of employment and shared economic well-being for all. The Government is now contemplating introducing a new *National Pharmaceutical Policy* with the following objectives:

- Making essential drugs accessible at affordable prices to the common masses;
- Providing a longer-term stable policy environment for the pharmaceutical sector;
- Making India sufficiently self-reliant in end-to-end indigenous drug manufacturing;
- Ensuring world-class quality of drugs for domestic consumption & exports;
- Creating an environment for R&D to produce innovator drugs;
- Ensuring growth and development of the Indian Pharmaceutical Industry.

National Pharmaceutical Pricing Authority (NPPA)

NPPA is an attached office of the Department of Pharmaceuticals and implements the National Pharmaceutical Pricing Policy 2012 and the Drugs Prices Control Order (DPCO) as amended from time to time which was issued by the Department. NPPA fixes ceiling price to all Drugs notified under Schedule-I of the DPCO, 2013 and monitors annual price increase for these medicines.

In February 2019, NPPA launched a Pilot for price regulation of 42 anti-cancer drugs as proof of concept for Trade Margin rationalization of non-scheduled drugs. So far, 526 brands have reported saving up to 90%, amounting to Rs. 984 crore per annum. It is estimated that NPPA has affected a saving of almost Rs.12, 447 crores to the consumers since its constitution.

For effective monitoring of DPCO in the country, such as monitoring of notified prices of medicines, detection of violation of the provisions of DPCO, pricing compliance, and ensuring availability of medicines, cooperation and involvement of State Drug Controllers (SDCs) is important. Because of that, NPPA has decided to set up Price Monitoring and Resource Units (PMRUs), as registered societies under the Societies Registration Act in all the 37 states/UTs to support NPPA and respective State Drug Controllers.

Patent Regime

(Adapted from the Book 'Innovation, Economic Development and Intellectual property in India and China by Kung-Chung Liu & Uday S.Racherla, 2019)

Before the 2005 amendment, the Indian Patents Act allowed the manufacturers to patent their manufacturing method for a given product. Before 2005, local pharmaceutical companies were allowed to reverse-engineer the best-selling drugs and produce cheap generic drugs for the domestic market and export them to Russia, China, Brazil, and Africa. In the process of producing generic drugs, Indian pharmaceutical companies accumulated extensive experience and trained their technical personnel. During the transition period, Indian domestic pharmaceutical companies allied with each other to conduct R&D and to manufacture pharmaceuticals for multinational pharmaceutical companies. The Indian domestic pharmaceutical industry grew rapidly, and some local pharmaceutical companies embarked on medicinal R&D and applied for drug patents in the United States and the European Union. The transition decade was the fastest growing period for the Indian pharmaceutical companies, and the industry became one of the most lucrative sectors in India. Research shows that the average profit margin (profit as a percentage of sales) of the pharmaceutical industry was about 8.8% in 1995 (while the profit margin of the chemical industry was 5.8%, that of the food and beverage industry was 4.8%, and that of the machinery manufacturing industry was 5.5%) and shot up to 15.4% in 2005. At the same time, generic drugs entering the Indian market grew rapidly. From 1980 to 1990, only 10 generic drugs entered the Indian market; the number reached 99 during the period 1990–1995, 156 during the period 1995–2000, and 262 during the period 2000–2005. It can be inferred that the transition period was a golden decade for the development of the Indian pharmaceutical industry. After 2005, India started to examine product patent applications, which led Indian pharmaceutical companies to adjust their product development policy and to increase R&D investment. At the same time, multinational pharmaceutical and biotech corporations started to increase contracted R&D and manufacturing in the Indian market.

Compulsory Licensing

In the Patents Act of 1970, there was already a special chapter for compulsory licenses, and the system was further improved in the Patents (Amendment) Act of 2002 and 2005. The compulsory licensing system creates more chances for voluntary licensing negotiation between the domestic Indian pharmaceutical companies and multinational corporations to succeed. According to Indian Patent Law, after the expiration of 3 years from the date of the sealing of a patent, any person interested may make an application to the Controller. The applicant is required to first attempt to procure a voluntary license from the patentee before applying for a compulsory license. If this attempt does not come to fruition within 6 months of the initial request, the applicant is entitled to file a compulsory license application. In 2012, the Controller, upon the application of Natco Pharma Ltd. (Natco), granted a compulsory license on Nexavar (Sorafenib Tosylate), a kidney cancer medicine patented by Bayer Corporation (Bayer). This is the first-ever and only compulsory license in India. In 2008, Bayer obtained the Indian patent (Patent No. 215758) on *sorafenib tosylate* and launched the drug into the Indian market under the same trade name, "Nexavar," selling it at Rs. 275,000 (US$ 5500) per patient per month in India. Bayer's drug was available only to a small percentage of eligible patients (slightly above 2%), which failed to meet the "reasonable requirements of the public." The 2012 price of Rs. 280,000 per month (approximately US$5600) to the patient does not meet the condition that the drug must be available to the public at a "reasonably affordable price." This is based on the purchasing power of patients in India. Natco was allowed to sell the drug to patients within India at a price of no more than Rs.8800 (approximately US$176) per month, which is 97% cheaper than Bayer's price. Natco was required to pay a 6% royalty to Bayer.

Pradhan Mantri Bharatiya Janaushadhi Pariyojana (PMBJP)

To make quality generic medicines available at affordable prices to all, Pradhan Mantri Bhartiya Jan Aushadhi Pariyojana (PMBJP) was launched in November 2008. Under this scheme, dedicated outlets known as Pradhan Mantri Bhartiya Janaushadhi Kendra (PMBJK) are opened to provide generic medicines. Under the Jan Aushadhi scheme, at least one Jan Aushadhi Store will be set up in each District of the country.

A medicine under PMBJP is priced on the principle of a maximum of 50% of the average price of the top three branded medicines. Therefore, the price of Jan Aushadhi Medicines is cheaper at least by 50% and in some cases, by 80% to 90% of the market price of branded medicines.

Uniform Code of Pharmaceuticals Marketing Practices (UCPMP)

It is a voluntary code issued by the Department of Pharmaceuticals related to marketing practices for Indian Pharmaceutical Companies as well as the Medical Devices Industry.

According to UCPMP companies or their associations/ representatives shall not extend any hospitality like hotel accommodation to healthcare practitioners and their family members under any pretext. The implied meaning of this is that even extending benefits to the doctors through associations is unethical.

Key features of UCPMP are as follows: -

- A drug can be promoted only after receipt of marketing approval by a competent authority and promotion should be consistent with marketing approval
- Claims for the usefulness of the drug must be based on an up-to-date evaluation of all the evidence
- The word "safe" must not be used without qualification and it must not be stated categorically that medicine has no side effects, toxic hazards, or risk of addiction
- The word "new" must not be used to describe any drug which has been promoted in India for more than 12 months
- Comparisons of drugs must be factual, fair, and capable of substantiation.
- Brand names of products of other companies must not be used in comparison unless prior consent of concerned companies has been obtained
- Promotional material such as mailings and journal advertisements must not be designed to disguise their real nature
- No gifts, pecuniary advantages, or benefits in kind may be supplied, offered or promised, to persons qualified to prescribe or supply drugs, by a pharmaceutical company or any of its agents.
- Companies or their associations/representatives or any person acting on their behalf shall not extend any travel facility inside the country or outside, including rail, air, ship, cruise tickets, paid vacations, etc., to Health Care Professionals and their family members for vacation or for attending the conference, seminars, workshops, CME program, etc.

- Free samples of drugs shall not be supplied to any person who is not qualified to prescribe such a product.
- UCPMP Code states that to appoint Medical Practitioners/HCPs as Affiliates there should be a written contract, a legitimate need for the services must be documented, and criteria for selecting affiliates must be directly related to the identified need.
- It also provides that the number of affiliates retained must not be greater than the number reasonably necessary to achieve the identified need and that the compensation must be reasonable and reflect the fair market value of the services provided.

SWOT Analysis of Indian Pharmaceutical Industry

Strengths

1. Cost Competitiveness (labor costs in India are up to 40% lower than other manufacturing hubs according to POBOS McKinsey's proprietary pharma operations benchmarking)
2. Developed Industry with Strong Manufacturing Base
3. Well Established R&D infrastructure
4. Access to a pool of highly trained scientists
5. Consistently growing domestic market
6. Strong marketing & distribution network
7. Competencies in Chemistry and process development
8. Large Patient base
9. Varied Disease profile

Weaknesses

1. Low investments in innovative R&D
2. Lack of resources to compete with MNCs for New Drug Discovery, Research, and commercialization of molecules on a worldwide basis
3. Lack of strong linkages between industry and academia
4. Low medical and healthcare expenditure in the country
5. Inadequate regulatory standards
6. Production of spurious and low-quality drugs
7. India continues to rely on imports of key starting materials and APIs (Active Pharma Ingredients) which exposes raw material supply disruptions and pricing volatility

Opportunities

1. Enhanced Universal Healthcare Coverage will improve access giving opportunity for the industry to penetrate the underserved market
2. Significant export potential (especially underpenetrated markets like Japan, China, etc.)
3. Patents for branded molecules with cumulative global sales of over USD 251 billion are expected to expire between 2018 and 2024, presenting opportunities in the generics market
4. (Evaluate Pharma World Preview 2018, Outlook to 2024)
5. Licensing deals with MNCs for NCEs (New Chemical Entities) and NDDS (New Drug Delivery System)
6. Marketing alliances for MNC products in domestic market and international market
7. Contract manufacturing arrangements with MNCs
8. Potential for developing India as a center for international clinical trials
9. A niche player in global pharmaceutical R&D
10. Growing new product classes like biosimilars, gene therapy, and specialty drugs
11. Opportunity to grow OTC market with increasing health awareness

Threats

1. Product patent regime poses a serious challenge to domestic industry unless it invests in research and development
2. R&D efforts of Indian pharmaceutical companies hampered by lack of enabling regulatory requirement
3. Drug Price Control Order puts unrealistic ceilings on product prices and profitability
4. Export effort hampered by procedural hurdles in India as well as non-tariff barriers imposed abroad
5. Increasing pricing pressure in a regulated market is squeezing margins and profitability

Impact of Covid-19 pandemic on Pharma Industry

The Covid-19 pandemic has severely impacted the pharma industry. The decline in pharma revenue and growth rate due to Covid 19 pandemic has already been discussed in the 'Indian Pharmaceutical Market Overview' section of this chapter. Some of the key factors impacting the industry are stated below.

1. **Patient access to Physician:** Due to lockdown, the doctors were not able to visit their clinics, and also patients were not able to visit the clinics or hospitals for non-covid treatment. Thus, there was a significant reduction in the number of patients adversely impacting the pharma business.

2. **Doctor visits by pharma companies:** Due to lockdown and social distancing norms, the face-to-face meetings by the pharma sales representatives were drastically reduced. This forced the pharma companies to innovate their physician engagement strategies which will be discussed in later chapters of this book.

3. **Supply chain disruption:** Lockdown imposed disruption in supply chain resulted in shortages of raw materials impacting manufacturing as well as the movement of finished goods resulting in shortages of essential medicines.

4. **Manufacturing of APIs:** India's dependence on China for API needs (70% imported from China) created a lot of hardships for pharmaceutical firms manufacturing certain key APIs. Some of the key APIs were crucial to mitigate the burden of accelerating diseases like tuberculosis, diabetics, and cardiovascular diseases in India.

   The Government of India has taken several important steps like incentive packages for the promotion of domestic manufacturing of critical key starting materials, drug intermediates, APIs, and medical devices, approval of pharmaceutical infrastructure development, providing tax exemptions and subsidies for the development and promotion of the pharmaceutical industry hubs, etc.

Future of Indian Pharmaceutical Industry

According to IPA (Indian Pharmaceutical alliance) report on Indian Pharmaceutical Industry – Way Forward, June 2019, the Indian pharmaceutical industry could grow to USD 120-130 billion by 2030 as depicted in the exhibit below.

Projected size of the Indian pharma industry, USD billion

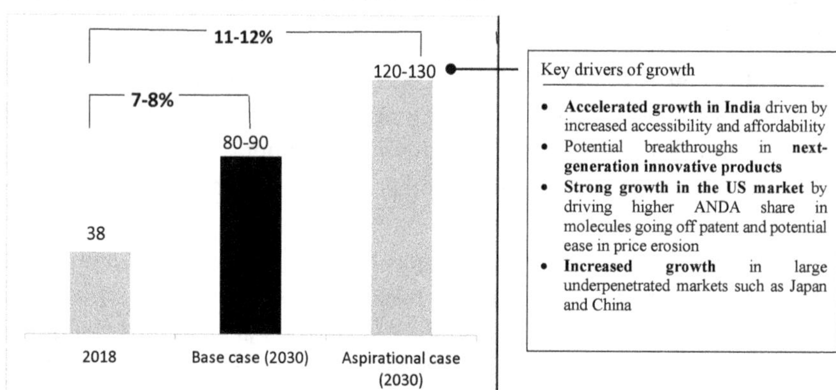

Fig. 2.5 Future of Indian Pharmaceutical Industry
Source: IQVIA, AIOCD, Pharmexcil, IPA Team analysis, secondary research

SUMMARY

This chapter deals with the Indian Pharmaceutical Industry landscape & an overview of the Indian Pharmaceutical Market. Key points discussed in this chapter are as follows:

1. Indian Pharmaceutical Industry is third largest in the World in terms of volume and tenth largest in terms of value. The total size of the industry (including drugs & medical devices) is around US$43 billion (Rs.3, 01,000 crores) and is currently having a growth rate of 7-8% in the drug sector and 15-16% in the medical device sector. Total exports (drugs and medical devices) are to the tune of US$20 billion (Rs.1, 47,420 crores) of which drugs form around 90% of the total exports. The imports amount to around Rs. 72,800 crores of which medical devices form around 52%.

2. Indian Pharmaceutical Market is valued at INR 145354 crores according to AIOCD AWACS MAT (Moving Annual Total) Dec 2020 data and has registered the growth of 3.1 % over the same period the previous year.

3. The market growth had dipped in 2017 due to demonetization and GST roll-out but picked up in 2018 and 2019. In the year 2020, the growth rate again dipped significantly due to the Covid-19 pandemic.

4. The top 20 companies contribute about 61.3 % to the total pharmaceutical market whereas the top 50 companies' contribution is about 82.8%.

5. Indian companies dominate the Indian Pharmaceutical Market with 80 % contribution whereas MNCs contribute 20 %.

6. The top 10 Therapeutic Categories contribute about 88% to the total pharmaceutical market.

7. Cardiac, Anti-infective, Gastrointestinal, Anti-Diabetic & Vitamins/Minerals/Nutrients are the Top 5 Therapeutic Categories and they contribute about 57% to the total pharmaceutical market.

8. Cardiac & Anti-Diabetic are the fast-growing Therapeutic Categories.

9. Mixtard is the biggest brand with a MAT value of INR 550.7 crores followed by Glycomet GP (547.7 crores), Lantus (542.9 crores), and Janumet (468.6 crores). All top 4 brands are from the Anti-Diabetic category.

10. Zincovit is the fastest-growing brand in the Top 20 brands, primarily due to the Covid-19 pandemic as zinc supplements were recommended for enhancing immunity.

11. The Drugs and Cosmetics Act 1940 and Rules 1945 regulate the import, manufacture, distribution, and sale of drugs and cosmetics. The Central Drugs Standard Control Organization (CDSCO) under the Directorate General of Health Services is the National Regulatory Authority (NRA) of India.

12. Under the Drugs and Cosmetics Act, CDSCO is responsible for approval of Drugs, the conduct of Clinical Trials, laying down the standards for Drugs, control over the quality of imported drugs in the country, and coordination of the activities of State Drug Control Organizations.

13. The Department of Pharmaceuticals notified the National Pharmaceutical Pricing Policy intending to put in place a regulatory framework for pricing of drugs to ensure availability of essential medicines at reasonable prices while providing sufficient opportunity for innovation and competition to support the growth of the industry.

14. Before 2005, local pharmaceutical companies were allowed to reverse-engineer the best-selling drugs and produce cheap generic drugs for the domestic market and export them. After 2005, India started to examine product patent applications, which led Indian pharmaceutical companies to adjust their product development policy and to increase R&D investment.

15. India's first-ever compulsory license was granted by the Patent Office on March 9, 2012, to Hyderabad-based Natco Pharma for the production of a generic version of Bayer's Nexavar, an anti-cancer

agent used in the treatment of liver and kidney cancer, which gave cancer patients easy access to the drug.

16. To make quality generic medicines available at affordable prices to all, Pradhan Mantri Bhartiya Jan Aushadhi Pariyojana (PMBJP) was launched in November 2008. Under this scheme, dedicated outlets known as Pradhan Mantri Bhartiya Janaushadhi Kendra (PMBJK) are opened to provide generic medicines.

17. The Uniform Code of Pharmaceuticals Marketing Practices (UCPMP) has laid down guidelines related to promotion, sampling, gifting, and other marketing practices.

18. According to UCPMP companies or their associations/ representatives shall not extend any hospitality like hotel accommodation to healthcare practitioners and their family members under any pretext. The implied meaning of this is that even extending benefits to the doctors through associations is unethical.

19. Strengths of the Indian Pharmaceutical Industry are cost competitiveness, strong manufacturing base, R&D infrastructure, highly trained scientists, growing domestic market, strong marketing & distribution network, competencies in process development, large patient base, and varied disease profile.

20. Weaknesses of the Indian Pharmaceutical Industry are low investments in innovative R&D, lack of resources to compete with MNCs for new drug discovery, lack of strong linkages between industry and academia, low healthcare expenditure, inadequate regulatory standards, production of spurious and low-quality drugs, dependence on imports of key starting materials and APIs which exposes raw material supply disruptions and pricing volatility.

21. Opportunities for Indian Pharmaceutical Industry are to penetrate the large underserved domestic market, significant export potential, opportunities in generics market due to patent expiry, licensing deals with MNCs for NCEs and NDDS, marketing alliances for MNC products in domestic market and international market, contract manufacturing, the potential for developing India as a Centre for international clinical trials, to be a niche player in global pharmaceutical R&D, growing new product classes like biosimilars, gene therapy, and specialty drugs & opportunity to grow OTC market with increasing health awareness.

22. Threats for Indian Pharmaceutical Industry are product patent regime poses a serious challenge to domestic industry unless it invests in research and development, R&D efforts of Indian pharmaceutical

companies hampered by lack of enabling regulatory requirement, Drug Price Control Order puts unrealistic ceilings on product prices and profitability, export effort hampered by procedural hurdles in India as well as non-tariff barriers imposed abroad & increasing pricing pressure in a regulated market is squeezing margins and profitability.

23. According to an IPA report on Indian Pharmaceutical Industry – Way Forward, June 2019, the Indian pharmaceutical industry could grow to USD 120-130 billion by 2030.

REFERENCES

1. Annual Report (2019-20), Government of India, Ministry of Chemicals & Fertilizers, Department of Pharmaceuticals.

2. EY-FICCI Indian Pharma Report, Feb 2021

3. https://www.india-briefing.com/news/indias-pharmaceutical-industry-investment-trends-opportunities-incentives-18300.html/ accessed on Oct 24, 2021

4. Indian life sciences: Vision 2030, FICCI June 2015, Growth estimated by IHS Markit.

5. Export-Import Data Bank, Department of Commerce, PHARMEXCIL, IDMA report on "Journey towards Pharma 2020 & beyond".

6. Fact sheet on FDI from April 2000 to June 2021, Department of Industrial Policy & Promotion available at https://dpiit.gov.in/sites/default/files/FDI_Factsheet_June2021.pdf accessed on Oct 3, 2021.

7. Press Information Bureau; "Affordable Efficacious Medicines – All Roads Leads to India", 2013 report by IDMA; Brandindiapharma.in.

8. Gx prescriptions account for 90% of prescriptions, Indian companies account for 40% share of Gx prescriptions, IQVIA 2019.

9. India Brand Equity Foundation, Indian Pharmaceuticals Industry Report, Aug 2021 available at https://www.ibef.org/industry/pharmaceutical-india.aspx accessed on Oct, 20,2020

10. Pharmaceuticals: India's Generics Flow into Africa, African Businesses Magazine, 19 January 2012 available at https://african.business/2012/01/economy/pharmaceuticals-indias-generics-flow-into-africa/ accessed on Aug 21, 2021

11. AIOCD AWACS Report, MAT Dec 2020 data.

12. https://ipapharma.org/portfolio/regulations-and-guidelines/ accessed on July 12, 2021.

13. Sawant AM, Mali DP, Bhagwat DA, Regulatory Requirements and Drug Approval Process in India, Europe, and the US, Pharmacet Reg Affairs, 2018, 7:210.

14. www.cdsco.gov.in

15. Book 'Innovation, Economic Development and Intellectual property in India and China by Kung-Chung Liu & Uday S.Racherla, 2019.

16. Circular on Uniform Code of Pharmaceuticals Marketing Practices (UCPMP) dated 12th Dec 2014 by Government of India, Ministry of Chemicals & Fertilizers, Department of Pharmaceuticals.

17. POBOS McKinsey's proprietary pharma operations benchmarking.

18. Evaluate Pharma World Preview 2018, Outlook to 2024

19. IPA (Indian Pharmaceutical alliance) report on Indian Pharmaceutical Industry – Way Forward, June 2019.

<h2>C H A P T E R 3</h2>

UNDERSTANDING PHARMACEUTICAL CUSTOMER

LEARNING OUTCOMES

After reading this chapter, you should be able to:

- Understand that pharmaceutical customers & consumers are different
- Understand the customer needs
- Get insight into models and theories of prescribing behavior
- Understand the Pyramid of Influence

Marketing starts with the customer & ends with the customer. What this implies is that the starting point of marketing is understanding of the customer – the needs, the buying behavior (in pharmaceutical marketing it would be prescribing behavior) & the endpoint is customer satisfaction leading to brand loyalty. The best marketing decisions are made when we understand our customers. Putting customers first is not new. We have enough evidence to prove that meeting customers' needs creates value and delivers a competitive advantage.

PHARMACEUTICAL CUSTOMER VS CONSUMER

Who is the customer for a Pharmaceutical Product? A 'Doctor' who prescribes the medicine is the customer. The patient who takes the medicine is the 'consumer'.

In Pharmaceutical marketing the customer and consumer are always different whereas in consumer marketing the customer and consumer could be the same or different. A doctor is a customer whereas a Patient is a consumer of pharmaceutical products (we are referring to prescription products here and not the OTC product). The patient invariably follows the doctor's prescription & has no role (or a very limited role) to play in prescription decision making. Whereas, in the case of a consumer product (e.g., soap), the customer and the consumer could be the same if the person buys the soap for himself/herself or could be different if buys soap for others. The difference between pharmaceutical marketing and consumer product marketing will be covered in detail in the next Chapter.

UNDERSTANDING CUSTOMER NEEDS

Let's take a pharmaceutical industry-related example to understand the concept of Need, Want & Demand. When a patient suffering from pain comes to a doctor, the 'need' is to give a 'pain reliever'. The doctor wants a pain reliever that is effective with fewer side effects. In case the patient is a well-informed, busy working executive, the doctor demands a pain reliever which is a new, strong, modern dosage form (e.g., a coated dosage form to avoid gastrointestinal side effects) from a reputed company.

The Doctor prescribes the drug based on his or her understanding of the patient's need and the knowledge of the drug best suited to address the need. Although the doctor is engaged in an educated search and decision-making activity, he or she may not be fully prepared to prescribe a product until the product details are formally communicated by pharmaceutical sales representatives, journal advertising, research conferences, and so on (Carter et al. 2006).

DOCTOR DECISION MAKING PROCESS

The Consumer Buying Decision Making process given by Engle Blackwel Model consists of five stages – Problem Recognition, Search for Alternatives, Evaluation, Purchase and Post Purchase Outcomes. Adapting this model to Doctor Prescription Decision Making Process, different stages could be stated as follows:

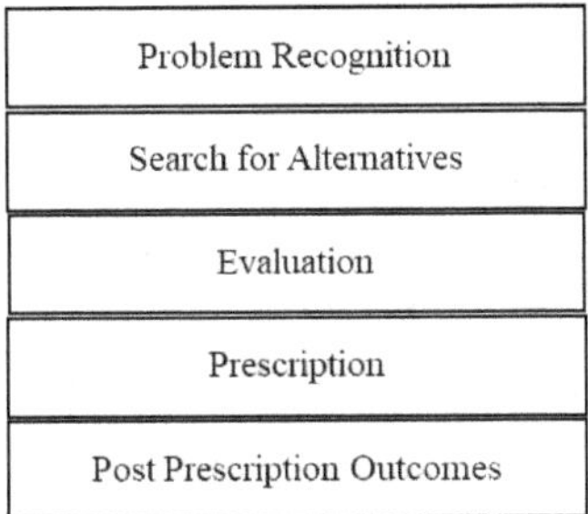

Fig. 3.1 Doctor Prescription Decision Making Process

Let's take an example. Suppose a patient comes to a doctor with a fever accompanied by cough and cold. The problem could be a respiratory tract infection. Alternatives would be various antibiotics used to treat respiratory tract infections. Suppose the doctor diagnoses an upper respiratory tract infection and chooses to prescribe a brand of Amoxycillin. An evaluation may be on basis of various factors like experience with a similar case, a brand from a reputed company, a recent visit by a medical representative of a company having an antibiotic to treat respiratory tract infection, price, availability of the brand, etc. The prescription would be based on the evaluation of various factors in the doctor's mind. Before boiling down to a particular brand, the doctor may consider a few alternatives called a 'consideration set' which in turn depend on the recall of choice alternatives from memory, known as 'retrieval set' (adaptation: Alba and Chattopadhyay, 1985). From this consideration set, the doctor chooses to prescribe a particular brand depending on brand image, company marketing that brand, medical representative visiting that doctor, price, promotional inputs, etc.

Thus, the prescription decision-making process in the example given above could be stated as follows:

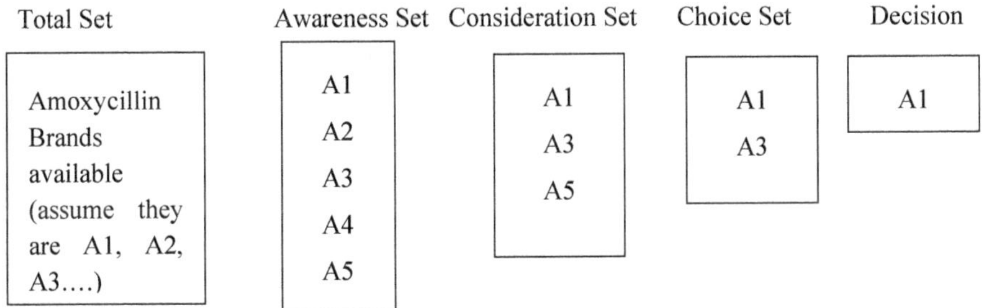

Fig. 3.2 Successive sets involved in Doctor Prescription Decision Making process

The pharma marketers need to ensure that their brand falls in the 'consideration set'.

There could be mainly two types of post prescription outcomes – i) the antibiotic brand is effective in treating the infection leading to the continued prescription of the brand in the future or ii) the antibiotic brand is ineffective in treating the infection resulting in discontinuing the prescription of that brand in future.

However, the prescription decision is a complex process. We will try to understand it through various research studies and models explained further in the chapter.

Doctor prescription decisions are not based on data alone. If doctors solely make decisions based on efficacy, safety, access, and cost, why wouldn't they just find information on the internet? It's because doctors are influenced by social and emotional reasons as well in making prescription decisions (Rehal Daniel, 2007).

"Physicians make decisions to improve patient care based upon influences that come from others around them (social), their personal experiences, the patient's persuasiveness and impact (emotional), and reasons of logic (pragmatic)" (Rehal Daniel, 2007).

"When it comes to physicians' prescription drug choice an asymmetric social interaction or "peer effect" arises because non-specialist physicians may rely on prominent physicians, the "opinion leaders," to help reduce the uncertainty about their prescription choices. The role of opinion leaders becomes most salient when changes occur in the therapeutic environment because these typically lead to increased uncertainty about drug efficacy among the non-specialist physicians" (Nair et.al.,2010).

Most often, the prescription of a particular pharmaceutical brand is a matter of habit termed as 'habit persistence'. A thorough understanding of this 'habit persistence' is not only important to understand the doctor's prescription decision, but also to understand the impact of various promotional tools. Persistent physicians are not only harder to influence but they are also less sensitive to marketing activities. This would also mean that, once a pharmaceutical company has captured a persistent physician, competing companies will have a very difficult time influencing the physician to switch. As a result, these physicians are very attractive: not only they are more likely to stay with the drugs that they start prescribing, but they are also more difficult to influence once the category matures (Janakiraman et.al.,2008).

Besides the pharmacological criteria in deciding which drug to prescribe, demographic influences like the gender of the doctors, the age of doctors, and the practice location of the doctors are also rated essential factors in the doctor's decision to prescribe (Khazzaka 2019).

NEW DRUG PRESCRIPTION

Doctors typically go through several phases before they are willing to prescribe any new drug: "aware but skeptical", "comfortable", "confident" and "proactive prescriber" (Wosinska, 2004).

Doctors are interested in learning about new drugs and for this, they are dependent on the detailing and sampling done by the medical representatives and also the behavior of other doctors. The adoption of the new drug by a doctor increases as more doctors near him/her adopt it (Manchanda et.al.,2008).

Three latent prescription-behavior states characterize physicians' dynamic prescription behavior towards new drugs (Montoya et.al. 2010). These three states are "inactive" state (no prescription state), "infrequent state" (somewhat more favorable prescription behavior) & "frequent" state (physicians frequently prescribe the drug to their patients).

MODELS AND THEORIES OF PRESCRIBING BEHAVIOR

Various research studies (Knapp et.al., 1972, Hemminki 1975, Raisch 1990, Gallan 2005, Singh 2008, Kyle et.al. 2008, Godin et. al. 2008, Stros et.al 2015, Ali et.al.2017, Mehralian et.al.,2017, Murshid et.al.,2017 and Rizwan et.al.,2020) have given different models for prescribing behavior and the factors responsible for a prescription decision can be summarized as follows:

1. Severity of disease
2. Benefits and side effects of the drug
3. Specialty of doctor
4. Number of years of practice
5. Promotion by the pharmaceutical company
6. Visits by Medical Representative (MR)
7. Product knowledge gained through detailing by MR
8. Relationship between Doctor & MR
9. Samples. Gifts, conference sponsorship provided by pharmaceutical companies

10. Doctor's Habit (past prescribing behavior)
11. Price of the drug
12. Continuing Medical Education (CME) conducted/supported by the pharmaceutical company
13. Key Opinion Leaders (KOLs)
14. Opinions of colleagues (peer influence)
15. Scientific data derived from clinical trials
16. Patient's presenting symptoms and doctor's diagnosis
17. Journal Advertisement
18. Doctor – pharmacist collaboration
19. Drug availability
20. Patient's request for the drug

THE PSYCHOLOGY OF RECIPROCITY

To understand this concept let's take an example of a pen gifted by a pharmaceutical company medical representative to the doctor. "The pen works as a marketing tool on two distinct levels. On the surface, it is a simple informational device, reminding the physician of the name of the particular drug, perhaps even reinforcing one of its attributes with a catchy slogan. But the effectiveness of this pen as a marketing tool lies in a much deeper psychological reality. This pen triggers the law of reciprocity. How do you say no to someone who has just given you a nice gift? These little gifts work, however, because they are gifts. No strings attached. And the lack of strings triggers reciprocity" (Ubel, 2009). Pharmaceutical companies understand the psychology of reciprocity and therein lies their power to influence medical practice.

IMPACT OF COVID 19 PANDEMIC

Face-to-face interaction with doctors through medical representatives has been the standard approach adopted by pharmaceutical companies. However, with the Covid-19 pandemic, face-to-face interaction with doctors was not possible. Also, patients could not visit the doctor's clinic. All these pushed for the rapid digitization of healthcare. There was the need for pharmaceutical companies to engage with doctors digitally and the doctors, in turn, had to engage with their patients through various digital platforms. Doctors have also become more digital-savvy, and therefore the prescription decision-making is also influenced by how pharmaceutical companies engaged with doctors digitally.

DOCTOR PRESCRIPTION DECISION MAKING MODEL

Based on the various factors discussed above, the Doctor Prescription Decision Making Model could be drawn as follows:

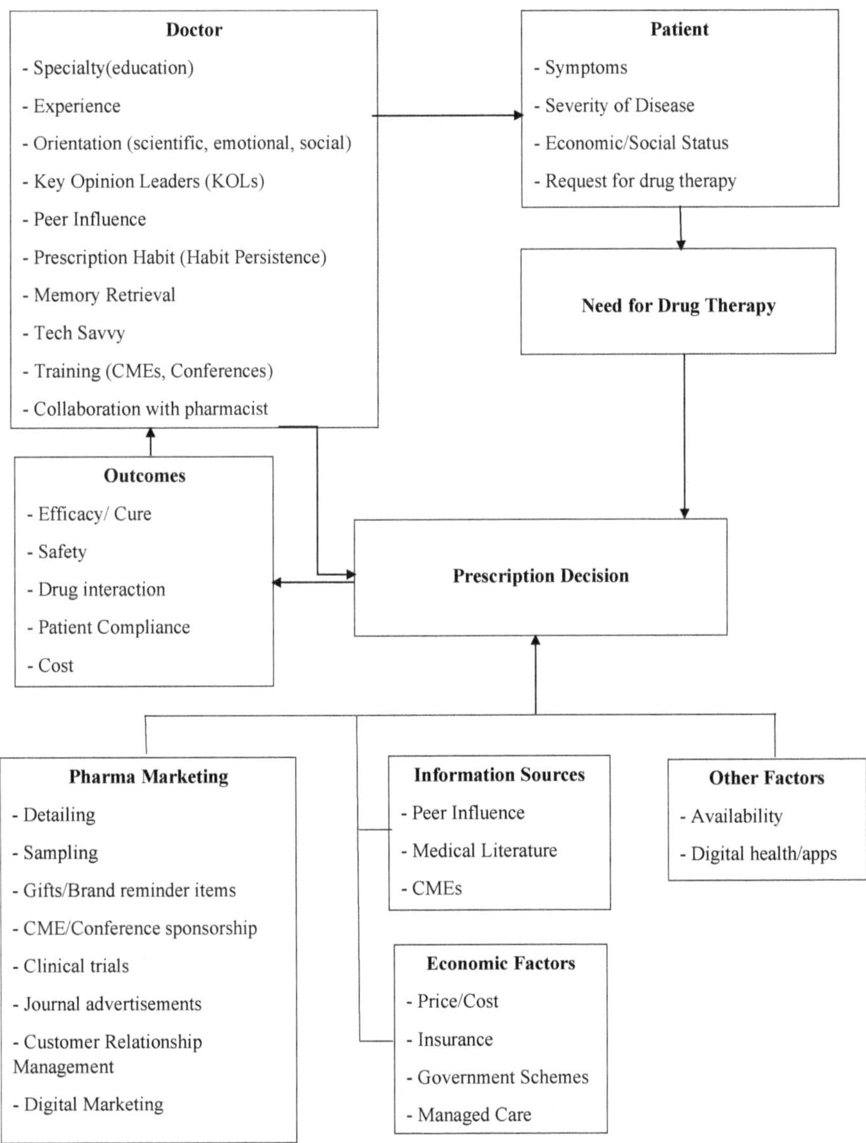

Fig. 3.3 Doctor Prescription Decision Making Model

PYRAMID OF INFLUENCE

We have seen above that the influencers in prescription decision-making could be the Key Opinion Leaders (KOLs), Peers, Medical Representatives, and Pharmacists. Based on this understanding the pyramid of influence could be drawn as follows: -

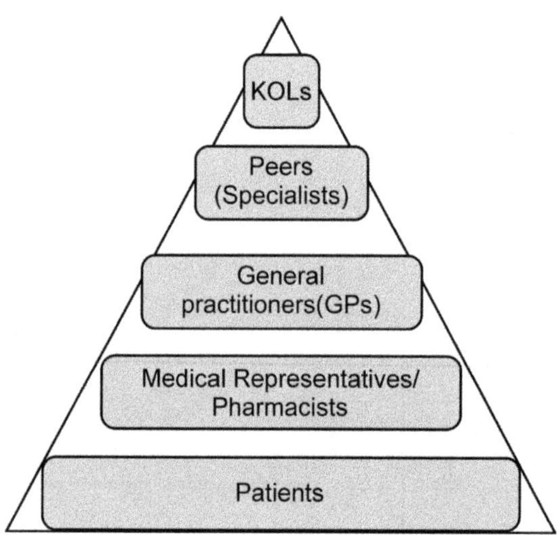

Fig. 3.4 Pyramid of Influence

SUMMARY

This chapter deals with the understanding of Pharmaceutical Customer which is the starting point of Pharmaceutical Marketing. Key points discussed in this chapter are as follows:

1. In pharmaceutical marketing customer and the consumer are different, the doctor is the customer whereas the patient is the consumer.

2. The doctor's decision-making process consists of five stages - Problem Recognition, Search for Alternatives, Evaluation, Prescription, and Post Prescription Outcomes.

3. The prescription would be based on the evaluation of various factors in the doctor's mind. Before boiling down to a particular brand, the doctor may consider a few alternatives called a 'consideration set' which in turn depend on the recall of choice alternatives from memory, known as 'retrieval set'.

4. Most often, the prescription of a particular pharmaceutical brand is a matter of habit termed as 'habit persistence'. A thorough understanding of this 'habit persistence' is not only important to understand the doctor's prescription decision, but also to understand the impact of various promotional tools. Persistent physicians are not only harder to influence but they are also less sensitive to marketing activities. This would also mean that, once a pharmaceutical company has captured a persistent physician, competing companies will have a very difficult time influencing the physician to switch.

5. Based on various research studies, the factors influencing prescription decision-making are the severity of disease, benefits and side effects of the drug, specialty of the doctor, years of practice, promotion by the pharmaceutical company, visits by Medical Representative (MR), product knowledge gained through detailing by MR, the relationship between the doctor & the MR, samples, gifts, conference sponsorship provided by pharmaceutical companies, doctor's habit (past prescribing behavior), price of the drug, Continuing Medical Education (CME) conducted/supported by the pharmaceutical company, opinions of colleagues (peer influence), scientific data derived from clinical trials, doctor's diagnosis, a journal advertisement, doctor – pharmacist collaboration, drug availability and patient's request for the drug.

6. Covid 19 pandemic pushed for rapid digitization of healthcare and therefore prescription decision-making is also influenced by how pharmaceutical companies engaged with doctors digitally.

REFERENCES

1. Carter Franklin J., Carol Motley, Alphonso Ogbuehi, and Jacqueline A. Williams, "An Application of a Repeat-Purchase Diffusion Model to the Pharmaceutical Industry," Advances in Business and Management Forecasting, Feb 2006, 4, 145–174.

2. J.F.Engel. R.D. Blackwell and P.W.Miniard, Consumer Behavior, International ed., Dryden Press, Florida, 1995.

3. Alba Joseph W. and Chattopadhyay Amitava, Effects of Context and Part-Category Cues on Recall of Competing Brands, Journal of Marketing Research,22(Aug 1985), 340-349.

4. Rehal Daniel, A prescription for change, supplement to Pharmaceutical Executive, March 2007, Vol VII, Section II.

5. Nair Harikesh S., Manchanda Puneet, Bhatia Tulikaa, Asymmetric Social Interactions in Prescription Behavior: The Role of Opinion Leaders, Journal of Marketing Research Vol. XLVII (October 2010), 883–895.

6. Janakiraman Ramkumar, Dutta Shantanu, Sismeiro Catarina, Stern Philip: Physicians' Persistence and Response to Promotions Management Science, 2008, 54(6), pp. 1080–1093.

7. Khazzaka Micheline, Pharmaceutical marketing strategies' influence on physicians' prescribing pattern in Lebanon: ethics, gifts, and samples, BMC Health Services Research, 2019, 19:80

8. Marta Wosinska, "PROPECIA: Helping Make Hair Loss History," HBS No.506-053(Boston: Harvard Business School,2004),7-8.

9. Manchanda Puneet, Xie Ying, and Youn Nara: The Role of Targeted Communication and Contagion in Product Adoption, Marketing Science,2008, 27(6), pp. 961–976.

10. Montoya Ricardo, Netzer Oded, and Jedidi Kamel: Dynamic Allocation of Pharmaceutical Detailing and Sampling, Marketing Science, 2010, 29(5), pp. 909–924

11. Knapp DE, Oeltjen PD, Benefits-to-risk ratio in physicians use when prescribing, Am. J. Public Health, 1972, 62(10):1346-1347

12. Hemminki E, Review of literature on the factors affecting drug prescribing, Soc. Sci. Med., 1975;9(2):111-116.

13. Raisch D W, A model of methods for influencing prescribing: part ii, a review of educational methods, theories of human inference and delineation of a model, DICP.1990;24(5):537-542

14. Gallan A, Factors that influence physicians 'prescribing of pharmaceuticals: a literature review, J Pharm., Mark, Manage, 2004;16(4):3-46

15. Singh R, Network connectedness of pharmaceutical sales rep (FLE) – physician dyad and physician prescription behavior: a conceptual model, J Med., Mark.,2008;8(3):257-268.

16. Kyle G J, Nissen L M, Tett S E, Pharmaceutical company influences on medication prescribing and their potential impact on quality use of medicines, J. Clin., Pharm., Ther., 2008;33(5):553-559.

17. Godin G, Belanger- Gravel A, Eccles M, Grimshaw J, Healthcare professionals' intentions and behaviors: a systematic review of studies based on social cognitive theories, Implement Sci.2008;3:36.

18. Stros M, Lee N, Marketing dimensions in the pharmaceutical industry: a systematic literature review, J Strategic Mark., 2015;23(4):318-336.

19. Ali Murshid M., Mohaidin Z., Models, and theories of prescribing decisions: A review and suggested a new model, Pharmacy Practice 2017, Apr-Jun: 15(2):990.

20. Mehralian, G., Sharif, Z., Yousefi, N., and Akhgari, M., Physicians' loyalty to branded medicines in low middle-income countries: A structural equation modeling. Journal of Generic Medicines, 2017, Vol. 13, No. 1, pp. 9-18.

21. Murshid, M.A., and Mohaidin, Z., Physicians' perceptions towards brand medicine and its effect on prescribing: A narrative review, Journal of Generic Medicines, 2017, Vol. 13, No. 4, pp. 157-183.

22. Ahmed Raheem Rizwan, Streimikiene Dalia, Abrhám Josef, Streimikis Justas, Vveinhardt Jolita, Social and Behavioral Theories and Physician's Prescription Behavior, Sustainability, 2020, 12(8), 3379.

23. Free Market Madness: Why Human Nature is at Odds with Economics and Why it Matters by Peter A. Ubel, Jan 2009.

PHARMACEUTICAL MARKETING – HOW IS IT DIFFERENT FROM CONSUMER MARKETING?

LEARNING OBJECTIVE

After reading this chapter, you should be able to:

- Understand how pharmaceutical marketing is different from marketing of consumer goods
- Learn strategies for marketing of consumer goods that can be adopted by pharmaceutical marketers

PHARMACEUTICAL MARKETING – HOW IS IT DIFFERENT FROM MARKETING OF CONSUMER GOODS

Pharmaceutical Marketing is unique and there are several differences between pharmaceutical marketing and marketing of consumer goods. The major differences are explained below.

1. **Customer vs Consumer:** In pharmaceutical marketing, the customer and consumer are always different whereas in consumer marketing the customer and consumer could be the same or different.

The doctor is the customer whereas the patient is the consumer for a pharmaceutical product. In the case of a consumer product like soap, the consumer and customer could be the same if the person buys the soap for himself/herself or could be different if the soap is bought for consumption of someone else.

2. **Customer Persona/Profile:** The pharmaceutical customer(doctor) is a highly educated, well-informed, sophisticated customer whereas a customer for consumer product may not be educated, well-informed or sophisticated. Therefore, the pharmaceutical product salesman (MR) needs to be knowledgeable, well-informed, and sophisticated which may not be the case in the case of a salesman of consumer products.

3. **High-involvement vs low-involvement:** The pharmaceutical product is bought on doctor's prescription which is high-involvement decision-making. A consumer product buying decision on the other hand could be low-involvement decision making, for e.g., there could be impulse purchase of chocolate when somebody is visiting a retail store. In other words pharmaceutical product is purchased only when there is an established need whereas a consumer product may be bought out of novelty, curiosity, trial, discount, etc.

4. **Knowledge-intensive:** The pharmaceutical industry runs on the most important wheel of knowledge. Perhaps no other industry is as knowledge-intensive as the pharmaceutical industry (Sultana et.al., 2009). Therefore knowledge management is very important in pharmaceutical marketing. This may not be the case while marketing a consumer good.

5. **Personal Selling vs Mass Marketing:** Pharmaceutical marketing focuses on 'personal selling' i.e., through face-to-face promotion(detailing) by the medical representatives to individual doctors.

   Consumer goods like soap on the other hand are promoted to the masses, e.g., TV advertising

6. **One-to-one vs one-to-many communication:** In the case of pharmaceutical marketing, the communication is one-to-one to individual doctors. The communication is to the doctor (influencer) and not to the end consumer, thus it is indirect communication. The

benefit of face-to-face communication is instant feedback from the doctor is possible.

In the case of consumer goods, the communication is one–to–many, e.g., when a TV advertisement of a consumer product is shown, the brand is exposed to many potential consumers at the same time. The brand is directly communicating to potential consumers simultaneously. Communication is a one-way communication only and no instant feedback from consumers is possible.

7. **Regulation:** The pharmaceutical industry is highly regulated. In India, the import, manufacture, distribution, and sale of drugs and cosmetics are regulated by the Drugs and Cosmetics Act (DCA) and its subordinate legislation, the Drugs and Cosmetics Rules (DCR). There are various regulatory bodies like MoHFW(Ministry of Health and Family Welfare), FDA (Food & Drug Administration), DPCO (Drugs Prices Control Order), DGCI(Drugs Controller General of India), CDSCO(Central Drugs Standard Control Organisation), etc.

Consumer goods also need to be of specified standards, but the rules are not as stringent as in the case of the pharmaceutical industry.

8. **Distribution:** Pharmaceutical products are sold only through registered chemist shops. For the sale and distribution of pharmaceutical products, a drug license is essential. Thus, access to pharmaceutical products is limited only to the universe of chemist outlets. With e-pharmacies, access has improved but is still limited to the digital-savvy population.

Consumer goods on the other hand are sold through varied retail outlets like grocers, departmental stores, chemist shops, pan-bidi shops, etc. and therefore are readily available.

9. **Known Customers vs Customer anonymity:** In Pharmaceutical marketing, the customer is known e.g., the company/medical representative knows the doctors who prescribe their product/s. Thus, the communication to a particular doctor can be modified depending upon his/her individual need.

In the case of consumer goods, the customers/consumers are not known. The communication is uniform to the entire target audience. With digital marketing targeted marketing is possible but limited to the digital-savvy population only.

The above points are summarized in the table (4.1) below.

Table 4.1 Pharmaceutical Marketing *vs* Consumer Goods Marketing

Sr.No.	Parameter	Pharmaceutical Marketing	Consumer Goods Marketing
1	Customer & Consumer	Both different	Could be the same or different
2	Customer profile	Educated, well informed, sophisticated	May not be educated, well informed, or sophisticated
3	Buying decision	High-involvement	Could be low-involvement or high-involvement
4	Knowledge	Knowledge-intensive	Not that knowledge-intensive
5	Promotion	Personal selling	Mass marketing
6	Communication	One-to-one	One-to-many
7	Regulation	Highly regulated	Not that stringent
8	Distribution	Selective, need a license	Intensive
9	Knowledge about customer	Known customer	Unknown customer

What can pharma marketers learn from consumer goods marketing?

Traditionally pharmaceutical marketing has been focusing on launching new innovative products, increasing doctor coverage, and managing relationships with doctors. "A major problem with the industry appears to lie in the constant cycle of product improvement; hence new brands are constantly being introduced at the expense of existing products. Thus, when more effective cancer treatments and a novel molecule for the treatment of schizophrenia arrive through the pipeline, existing brands become demoted in importance and transferred to cash cow status. It is hardly surprising then that, with support being withdrawn, brands that were carefully nurtured and polished in their youth become old and decay with alarming rapidity"(Blackett et.al.,2001). Secondly, it has become much more difficult to identify blockbuster drugs and it is costly as well(Schuiling et.al., 2004). Thirdly, the increase in doctor coverage through increased sales force can reach saturation level and a further increase in salesforce may not give incremental benefits.

Consumer goods companies, on the other hand, focus on brand building. Not much attention has been given to building brands by pharmaceutical companies. Therefore, let's learn from consumer goods companies about their brand-building strategies.

1. **Brand management organization:** Consumer goods companies dedicate a lot of management attention and effort to managing their brands. Branding is a strategic priority and the marketing team works closely with Research & Development right from the beginning of the product development process(Schuiling et.al., 2004). The focus is on maximizing long-term brand growth.

   In the pharmaceutical industry, the marketing team is often involved late in the product development process. Also, pharmaceutical marketing often is sales-driven rather than focusing on strategic marketing. Few big pharmaceutical companies are exceptions as they have started focusing on marketing-driven strategies. Thus creating a culture where the importance of branding is understood is critical. No amount of marketing-driven orientation is going to help unless it is implicitly accepted that branding is of commercial value and can help optimize scientific success (Blackett et.al.,2001).

2. **Branding strategies:** The trend in the consumer goods industry is to associate a new product with a well-known big brand or corporate brand name to benefit from existing awareness and a strong image(Schuiling et.al., 2004).

   These involve various strategies like brand name strategies, brand architecture strategies, brand extension strategies, cobranding strategies, etc. which will be covered in detail in Chapter 13.

3. **Building brand values:** At the heart of all brands lies a set of values. Values are beliefs that customers have about a brand that they find intuitively attractive, and which are likely to influence their purchase decision. As such they represent the foundation stones of the buyer-seller relationship. These beliefs can sometimes be entirely rational and based on the functionality of the product per se emotional. Consumer goods companies focus on building emotional values for their brands. They focus on an in-depth understanding of consumers by adopting various research techniques. The strategies are based on this understanding of consumer insights. The pharmaceutical industry on the other hand focus on functional attributes and benefits of the product and rarely research is carried out to understand the customer insights. If we examine any leading therapeutic segment we will find that choice is abundant in terms of products and types of treatment. For example, in the treatment of hypertension, there is a range of calcium antagonists, including amlodipine, felodipine, isradipine, lacidipine, nicardipine, nifedipine, nisoldipine, etc. However, the degree to which any of these can claim to have developed competitive

differentiation is questionable; their advantages have been relatively short-lived and have relied mainly on patents(Blackett et.al.,2001).

If a brand is a body of cultural signification, getting that signification ' right 'is critical. It requires a clear understanding of customers 'thoughts and feelings about treating the relevant group of patients, their thoughts and feelings about their role as a healthcare professional, and the brand choices they face when making a product decision (Kelly and Rupert,2009). While it is important to understand the functional approach to understanding doctors' attitudes toward medical treatments and products, it is equally important to adopt qualitative and quantitative research techniques that can be used to assess decision-makers 'emotional and non-rational thought processes. "By mining insights, pharma marketers can uncover the rational and emotional needs that exist in a category or product"(Pharmaceutical Executive Supplement, March 2004).

4. **Brand building strategies:** Consumer goods companies focus on long-term brand building and can leverage the power of brands. In the consumer goods industry, the role of brands as a source of competitive advantage has been well developed. Coca-Cola, Pepsi, Bisleri, Cadbury, Colgate which are leading consumer brands, have created a distinct identity for themselves through a consistent branding exercise. The power of branding in some instances has been so rewarding that the Colgate brand, for example, is often used as a generic term when associated with toothpaste.

   According to Kapferer (2001), "a product gives a certain efficacy while a brand gives more trust." Pharmaceutical brands engender trust, which is one of the most powerful emotions. As opposed to the effectiveness of cola drinks, relief from pain or illness is a far higher state of need than the alleviation of thirst. Hence trust in pharmaceutical brands is a vital component and once gained can be leveraged successfully through the brand. This trust built up with the brand also provides emotional justification(Blackett et.al.,2001).

   Pharmaceutical companies need to build their brand reputation and brand loyalty through pre-diagnosis relationships with the consumer. "Instead of creating messaging around how the product addresses the disease or diagnosis, brands need to start building relationships long before diagnosis occurs" (Robinson, PharmaVOICE,2014).

   "In the pharmaceutical industry, where the patent life and the duration of market exclusivity are limited, brands must work harder and faster to create lasting value that extends beyond a particular product's expected lifespan. It's about a brand name and a perceived value that resonate with prescribers and consumers alike, and the first step to achieving that is establishing a solid foundation of communication strategies that focus on educating and empowering these audiences and increasing their awareness and ability to make informed and enlightened decisions" (Robinson, PharmaVOICE,2014).

SUMMARY

This chapter deals with the understanding of how pharmaceutical marketing is different from marketing consumer goods and the consumer goods marketing strategies that can be adopted by pharmaceutical marketers. Key points discussed in this chapter are as follows:

1. Pharmaceutical marketing is different from consumer goods marketing. In pharmaceutical marketing customers and consumers are different, it is high-involvement, is knowledge-intensive, the customer i.e.doctor is known, is educated, well informed, and sophisticated which may not be the case in consumer goods marketing. Pharmaceutical marketing is highly regulated, distribution is selective and promotion is mainly through personal selling.

2. Pharmaceutical marketing mainly focuses on launching new innovative products, increasing doctor coverage, and managing relationships with doctors, and less on brand building.

3. Pharma industry focus has been in the constant cycle of product improvement; hence new brands are constantly being introduced at the expense of existing products. However, developing blockbuster drugs is time-consuming and costly.

4. Pharma companies can learn from consumer goods companies about their brand management organization, branding strategies, and brand building strategies.

REFERENCES

1. Sultana Sadika M and Manivannan, Knowledge Management in Pharmaceutical Marketing: Theoretical Models, The Icfai University Journal of Knowledge Management, Vol.VII, No.2, 2009, pg 62-76.

2. Blackett Tom, Harrison Tom, Brand Medicine: Use and future potential of branding in pharmaceutical markets, International Journal of Medical Marketing, Sep 2001, Vol.2, 1, 33-49.
3. Schuiling Isabelle, Moss Giles, How different are branding strategies in the pharmaceutical industry and the fast-moving consumer goods sector?, Brand Management, 2004, Vol.11, No.5, 366-380.
4. Kelly Donna and Rupert Edwin, Professional emotions and persuasion: Tapping non-rational drivers in health-care market research, Journal of Medical Marketing 2009, Vol.9,1, 3 – 9.
5. Pharmaceutical Executive Supplement, March 2004, Vol. I, Section III: Branding.
6. Robinson Robin, Brand identity in a consumer-driven world, PharmaVOICE, Sep 2014,20-23.
7. Robinson Robin, Brands cannot live on efficacy alone, PharmaVOICE, Sep 2014,26-28.

CHAPTER 5

PHARMA 4.0

LEARNING OUTCOMES

After reading this chapter, you should be able to:

- Understand what is Industry 4.0 & Pharma 4.0
- Understand emerging technologies and their application
- Understand the concept of patient centricity
- Understand how to track patient journey
- Get insight into patient-centered apps

INDUSTRY 4.0

The first industrial revolution spanned from about 1760 to around 1840 characterized by the construction of railroads and the invention of the steam engine, leading to mechanical production. The second industrial revolution, which was in the late 19th century and into the early 20th century, made mass production possible, with the advent of electricity and the assembly line. The third industrial revolution that began in the 1960s led to the digital revolution as it catalyzed the development of semiconductors, mainframe computing (1960s), personal computing (1970s and 80s), and the internet (1990s) (Klaus Schwab, World Economic Forum, 2016).

The fourth industrial revolution began at the turn of the 21st century and builds on the digital revolution. Characterized by "smart factories", the fourth industrial revolution enables the customization of products and the creation of new operating models. The fourth industrial revolution is not only about smart and connected machines and systems but its scope is much wider. Occurring simultaneously are waves of further breakthroughs in areas ranging from gene sequencing to nanotechnology, from renewables to quantum computing. It is the fusion of these technologies and their interaction across the physical, digital and biological domains that make the fourth industrial revolution fundamentally different from previous revolutions. The fourth industrial revolution has made possible new products and services that increase at virtually no cost the efficiency of our personal lives as consumers. Ordering a cab, finding a flight, buying a product, making a payment, listening to music, or watching a film – any of these tasks can now be done remotely. The internet, the smartphone, and the thousands of apps are making our lives easier, and – on the whole – more productive (Klaus Schwab, World Economic Forum, 2016).

PHARMA 4.0

Many health challenges, from heart disease to cancer, have a genetic component. Because of this, the ability to determine our genetic make-up efficiently and cost-effectively (through sequencing machines used in routine diagnostics) will revolutionize personalized and effective healthcare. Informed by a tumor's genetic make-up, doctors will be able to make decisions about a patient's cancer treatment. While our understanding of the links between genetic markers and disease is still poor, increasing amounts of data will make precision medicine possible, enabling the development of highly targeted therapies to improve treatment outcomes. Already, IBM's Watson supercomputer system can help recommend, in just a few minutes, personalized treatments for cancer patients by comparing the histories of disease and treatment, scans, and genetic data against the (almost) complete universe of up-to-date medical knowledge. 3D manufacturing will be combined with gene editing to produce living tissues for tissue repair and regeneration – a process called bioprinting. This has already been used to generate skin, bone, heart, and vascular tissue. Eventually, printed liver-cell layers will be used to create transplant organs. We are developing new ways to embed and employ devices that monitor our activity levels and blood chemistry, and how all of this links to well-being, mental health, and productivity at home and work.

Industry 4.0 which includes emerging technologies from connectivity to advanced analytics, robotics, and automation has the potential to revolutionize every element of pharma manufacturing, increasing productivity and reducing overall quality-control costs. Digitization and automation will also ensure better quality and compliance by reducing manual errors and variability, as well as allowing faster and effective resolution of problems (Han Yan et.al. McKinsey & Company, 2019).

Digital technologies such as robotics, data analytics, artificial intelligence, and the industrial Internet of things (IIoT) can deliver greater efficiency and also help in designing new drugs faster by using artificial intelligence to screen compounds and to increase responsiveness to customers (Sandle Tim, 2019).

Emerging Technologies - definitions and their application

Let's understand what are the key emerging technologies in healthcare, their definitions, and their application in brief. More detailed applications of emerging technologies will be discussed in later chapters.

Following are the key technologies and their definitions adapted from the book 'Tech Trends in Practice - 25 Technologies That Are Driving The 4th Industrial Revolution,2020' by Marr Bernard.

1. **AI, ML:** AI involves applying an algorithm (a rule or calculation) to data to solve problems, identify patterns, decide what to do next, and maybe even predict future outcomes. Machine learning is a subdiscipline of AI, and it involves creating machines (computers, smartphones, software, industrial equipment, robots, vehicles, etc.) that can learn.

   AI and ML tools can be used for predicting physician prescription behavior, the marketing channels they are most likely to respond to, and the messaging that will have the biggest impact. These tools also can be effective for engagement with patients.

2. **IoT, IoMT:** Smart devices, gadgets, or machines gather information and communicate that data via the internet – an example being your fitness tracker automatically sending activity data to an app on your phone.

   The IoT is poised to transform the healthcare industry, giving rise to its name: the IoMT (Internet of Medical Things). These IoMT devices can be used to help monitor patients, inform caregivers in the event of an emergency, and provide healthcare professionals with data that could inform diagnosis and ensure patients follow doctors' orders. Pharma marketers can leverage this for patient engagement and engagement with physicians.

3. **Big Data & Augmented Analytics:** "Big Data" refers to the exponential explosion in the amount of data being generated in this

digital age, while "augmented analytics" refers to the ability to automatically work with and generate insights from data.

Big data and augmented analytics can be used for optimizing pricing, forecasting demand, boosting productivity, strengthening the supply chain, and many other areas of the business to make improvements, generate efficiencies, save money, automate processes, etc.

4. **VR, AR, and MR:** Virtual reality (VR) means using computer technology to fully immerse the user in a simulated digital environment, to the extent that they feel like they're physically in that environment.

   Augmented reality (AR) is rooted in the real world, not a simulated digital environment. With AR, information or images can be overlaid onto what the user is seeing in the real world.

   Mixed reality (MR) is an extension of AR that brings the virtual and real worlds together, creating a more connected experience in which the virtual and real elements can interact.

   VR, AR, MR can be used for communicating with doctors or for conducting training sessions.

5. **Natural Language Processing:** Natural language processing (or NLP for short) refers to the technology that allows computers to understand human language.

   NLP can be used to improve healthcare and deliver better patient outcomes. McKinsey used NLP to enhance clinical benchmarking guidelines by analyzing clinical guidelines from multiple sources, then automatically organizing and classifying the information – resulting in a 60% decrease in the time needed to create clinical guidelines (www.mckinsey.com).

6. **Voice Interfaces and Chatbots:** Voice interfaces and chatbots are computer programs that allow humans to converse and interact with computers through either spoken commands or written text. Chatbots can be used to respond to medical questions, concerns, and symptoms.

7. **Computer Vision and Facial Recognition:** Computer vision is where machines (including computers, software, and algorithms) can "see" and interpret the world around them with facial recognition.

   In the healthcare sector, a staggering 90% of all medical data is image-based (www.healthcareitnews.com), meaning there are many valuable uses for computer vision. Microsoft's InnerEye software is one such example. The system can analyze X-ray images and identify possible tumors and other anomalies (www.microsoft.com).

8. **Robots & Cobots:** Robots can be defined as intelligent machines that can understand and respond to their environment and perform routine or complex tasks autonomously.

   Robots in Healthcare Robots are slowly beginning to change the face of medicine and augment the work of human healthcare professionals for example in conducting surgical procedures.

9. **5G and Faster Smarter Networks:** 5G is the fifth generation of cellular network technology, which together with other network innovations will give us much faster and more stable wireless networking, as well as the ability to connect more and more devices and enable richer, more varied streams of data.

   5G networks are playing a part in telemedicine, allowing doctors to examine patients and, soon, carry out surgery from hundreds of miles away.

10. **3D Printing:** 3D printing means creating a 3D object from a digital file by building it layer by layer. For example 3D printing of human tissue.

PATIENT CENTRICITY

Patients are becoming increasingly connected via mobile technology and social media. Nearly 50 percent of the Indian population use the internet regularly and smartphone penetration has reached 54 percent in the year 2020 (www.statista.com). This increased connectivity will have a profound impact on how the pharma industry and other key stakeholders will engage with patients.

Patient centricity can be defined as 'Putting the patient first in an open and sustained engagement of the patient to respectfully and compassionately achieve the best experience and outcome for that person and their family'(Yeoman G, et al. BMJ Innov 2017;3:76–83).

To be patient-centered means to understand the entirety of the situation a patient is facing, not just their disease. Many additional factors can impact that patient's well-being. Being patient-centric means learning and doing something for patients along the product life cycle and the patient journey.

What do people need?

- Connections with like-minded people
- Personalized content and updates
- Focused information and support from professionals

Mapping the patient journey can help identify the potential obstacles that may impact a patient's ability to receive the care his or her physician has prescribed and identify the best times to communicate and engage.

Being patient-centric for the pharma industry means ensuring that patients are front and center in all that we do — that today means boldly ushering in a new age of automation and digitization to ensure controls like we never had before on our products, processes, and quality systems (Gupta Nilesh, The Hindu Business Line, March 28, 2021).

How to track patient journey

Patients are the most important element of any pharma company's core mission. **Putting the patient at the center requires understanding the whole person, not just his or her illness.**

A patient's journey is unique, complex, and multi-dimensional reflecting tests, diagnoses, treatments, emotions, and behaviors that change over time.

Understanding their journeys and providing patients with support at each stage of treatment (i.e., hub enrollment, onboarding, and adherence) is essential in creating a valuable brand experience and a high-performing, high-impact program. By leveraging deep insights into access, affordability, and adherence barriers and how corrective actions improve adherence manufacturers can better serve patient needs with the right touchpoints at the right times. Through the power of predictive analytics and machine learning, manufacturers can rapidly identify patient behaviors and patterns to develop personas and predict the next best action for personalized engagement across direct, digital, and telehealth channels with just-in-time recommendations (PharmaVOICE, Oct 2021).

Tracking a patient's journey gives pharmaceutical marketers a tool that shows how to best help patients, caregivers, and physicians attain the promise of better health. Once we understand the patient mindset at each stage of the journey and the decisions that patients must make when moving from one phase to the next, we can tailor communications to influence their decisions.

Mapping a patient journey is a multi-step, iterative process reliant on robust data, clinical expertise, and analytics. The first step is to define the anchor codes for the journey in question such as a specific procedure, diagnosis, or prescription and the timeframe for analysis, which can range from a few days in the case of a course of antibiotics, to years for a disease such as diabetes. The next step is to identify clinically relevant claims to include in the analysis, which can be done bottom-up or top-down. Key metrics for analysis can then be defined. To do all of this, an expansive and longitudinal set of medical and pharmacy claims should be supplemented by other data, such as clinical data and social determinants of health. Especially for journeys that are complicated and/or less well understood, this process can be time-consuming; tools such as AI and natural language processing can help to automate and make the process more scalable.

The journey may begin with the patient's first and introductory post in an online healthcare forum: in a paragraph or two, patients typically tell the most relevant version of their story to orient the audience, making them able to provide appropriate help, information, and support. These highly efficient, mini online journey posts usually start pre-diagnosis, touch on multiple medical and treatments milestones, before ending up with a description of their current situation and their specific need and reason for posting within the forum. This all plays out throughout sentences, not pages, and these types of informative narratives communicate not only what happened, but also, often, how the patient felt throughout it all.

Brands can leverage patient opinion leaders to get their message directly to other patients in meaningful ways, for example in branded materials for websites, television, podcasts, and social media. "Patients' voices can also be leveraged as part of the training provided to healthcare professionals, sales teams, and other stakeholders.

PATIENT-CENTERED APPS

With the growing demand for a patient-centric healthcare ecosystem, app developers, mobile vendors, healthcare professionals, and the pharma industry are coming together to offer patient-centric healthcare apps. Patient-centric apps are defined as programs that offer health-related services like telemedicine, online video consultation, medical records storage, test reports, etc. In the case of management of diabetes, patient-centric apps offer services like tracking blood glucose measurements, maintaining a nutrition database, carbohydrate tracking, physical activity, and weight tracking.

Among the health condition management apps, the largest categories continue to focus on chronic conditions such as mental health and behavioral disorders, which account for 22%, followed by diabetes (15%) and heart and circulatory system apps (10%)(IQVIA report on Digital Health Trends, July 2021).

Pharma companies are using these digital health tools to augment the value of their products. Some pharma companies have launched simple companion apps, while others have made investments in digital therapeutics intended to treat conditions. E.g., Click Therapeutics and Boehringer Ingelheim agreement in the schizophrenia space to jointly develop and commercialize a mobile app for patients with schizophrenia & AstraZeneca's collaboration with Kardia to help patients with chronic kidney disease.

SUMMARY

This chapter deals with the understanding of concepts of Industry 4.0, Pharma 4.0, patient centricity, tracking patient journey, patient-centered apps, and emerging technologies and their application. Key points discussed in this chapter are as follows:

1. The fourth industrial revolution builds on the digital revolution. Characterized by "smart factories", the fourth industrial revolution enables the customization of products and the creation of new operating models. The fourth industrial revolution is not only about smart and connected machines and systems but its scope is much wider.

2. Industry 4.0 which includes emerging technologies from connectivity to advanced analytics, robotics, and automation has the potential to revolutionize every element of pharma manufacturing, increasing productivity and reducing overall quality-control costs.

3. Digital technologies such as robotics, data analytics, artificial intelligence, and the industrial Internet of things (IIoT) can deliver greater efficiency and also help in designing new drugs faster by using artificial intelligence to screen compounds and to increase responsiveness to customers.

4. Emerging technologies in Pharma are AI, ML, IoT, IoMT, Big Data & Analytics, VR, AR, MR, Chatbots, NLP, 5G, etc. These technologies have varied applications in predicting physician behavior, developing strategies for physician engagement, patient engagement, telemedicine, etc.

5. To be patient-centered means to understand the entirety of the situation a patient is facing, not just their disease. Many additional factors can impact that patient's well-being. Being patient-centric means learning and doing something for patients along the product life cycle and the patient journey.

6. Tracking a patient's journey gives pharmaceutical marketers a tool that shows how to best help patients, caregivers, and physicians attain the promise of better health. Once we understand the patient mindset at each stage of the journey and the decisions that patients must make when moving from one phase to the next, we can tailor communications to influence their decisions.

7. Patient-centric apps are defined as programs that offer health-related services like telemedicine, online video consultation, medical records storage, test reports, etc. Pharma companies are using these digital health tools to augment the value of their products either by launching simple companion apps or investing in digital therapeutics intended for treating certain conditions.

REFERENCES

1. The Fourth Industrial Revolution, Klaus Schwab, World Economic Forum, 2016

2. Han Yan, Makarova Evgeniya, Ringel Matthias, and Telpis Vanya, Digitization, automation, and online testing: The future of pharma quality control, McKinsey & Company, 2019

3. Sandle Tim, Becoming Pharma 4.0: How Digital Transformation Is Reshaping Pharmaceuticals, Aug. 27, 2019, https://www.biopharmatrend.com/post/109-becoming-pharma-40-how-digital-transformation-is-reshaping-pharmaceuticals/ accessed on July 2, 2021.

4. Marr Bernard, Tech Trends in Practice - 25 Technologies That Are Driving The 4th Industrial Revolution,2020.

5. Natural language processing in healthcare: www.mckinsey.com/industries/healthcare-systems-and-services/our-insights/natural-language-processing-in-healthcare

6. IBM Watson Health, Merge launch new personalized imaging tools at RSNA: www.healthcareitnews.com/news/ibm-watson-health-mergelaunch-new-personalized-imaging-tools-rsna

7. Project InnerEye – Medical Imaging AI to Empower Clinicians: www.microsoft.com/en-us/research/project/medical-image-analysis/

8. Internet and smartphone penetration in India, www.statista.com accessed on Oct 21, 2021

9. Yeoman G, Furlong P, Seres M, et al. Defining patient centricity with patients for patients and caregivers: a collaborative endeavor, BMJ Innov 2017;3:76–83.

10. Gupta Nilesh, Patient-centricity must drive the pharma sector, The Hindu Business Line, March 28, 2021

11. Kirsch Maria and O'Bryon Bill, Bridging the Patient Support Gap With Integrated Technology Solutions and Strategies, PharmaVOICE, Oct 2021

12. IQVIA report on Digital Health Trends, July 2021

13. Muoio D. Boehringer Ingelheim, Click Therapeutics ink $500M+ digital therapeutic development, commercialization deal. mobihealthnews. 2020 Sep 11. Available from: https://www.mobihealthnews.com/news/boehringer-ingelheim-clicktherapeutics-ink-500m-digital-therapeutics-development

14. Diagnostic and Interventional Cardiology. AliveCor collaborates with AstraZeneca to develop noninvasive potassium monitoring solutions. 2021 Mar 3. Available from: https://www.dicardiology.com/content/alivecor-collaborates-astrazeneca-developnoninvasive-potassium-monitoring-solutions

C H A P T E R **6**

MARKETING RESEARCH AND MIS

LEARNING OUTCOMES

After reading this chapter, you should be able to:

- Understand the role of Marketing Information System (MIS)
- Understand Pharma Databases –
    - Retail audit (AIOCD AWACS, IQVIA)
    - Prescription audit (SMSRC)
- Understand the Marketing Research process & its application

MARKETING INFORMATION SYSTEM (MIS)

Indian pharmaceutical industry is a fast-growing, dynamic industry, and therefore efficient and effective marketing decision making is important to stay ahead in the marketplace. There is a need to continuously monitor the changes that are taking place in the market, competition, physician prescribing behavior, patient behavior and trade behavior, internal sales trends, etc. It is important to collect, sort, analyze, evaluate and distribute the data for timely decision making. This is done through the Marketing Information System (MIS) which consists of-

- Internal records - sales, invoices, inventory, price, cost, receivables, payables
- Pharma databases - tracking sales of pharmaceutical products and doctors' prescriptions

- Market Intelligence - gathered through salesforce reports, trade partners, government databases, pharma databases
- Marketing Research - will be dealt with in detail in this chapter
- Decision Support System - comprises of technology, processes, analytics

PHARMA DATABASES

Two main types of pharma databases play an important role in marketing decision-making. These are as follows:

- Retail audit (AIOCD AWACS, IQVIA)
- Prescription audit (SMSRC)

Let's briefly try to understand each of these databases.

1. **AIOCD-AWACS:** AIOCD (All India Origin of Chemist and Distributors Ltd.) AWACS (Advance Warning Action & Correction System) is a joint venture between AIOCD & Trikaal Medi Info Tech Pvt Ltd. It's a pharma research company that tracks the secondary sales data of pharmaceutical products. The methodology followed for tracking is as follows: -

   - *Methodology:* They track "stockist out" data corresponding to secondary sales that are captured by electronic method. This is for total sales – hospital as well as trade sales. The audit takes total units sold into account and arrives at value after multiplying with a standard price to retailer (PTR). They follow European Pharmaceutical Market Research Association (EphMRA) classification for reporting with more sub-groups to suit the Indian market. Data from Jan 2015 onwards are available to all its clients.

   Salient Features are as follows: -
   - No of stockists empaneled: 9000
   - Frequency of reporting: Monthly, by 10th of the subsequent month.
   - They track 914 plus companies & 98,704 packs.
   - The online report is available which provides quick & easy access to data.
   - Provides data on closing Stock levels of competitor brands, categories & companies for 31 geographies.

- Higher and growing accuracy as they capture 65% through sample and project 35%.

- Instant New Product Capture - Any product where there is primary billing before the 20th of the month is captured in the Audit of that month itself.

- Highly accurate pack & price depiction.

- Canned reports – ready-made standard formats are emailed.

- Bonus offer as online tool query – by the pack, by category, etc. Company Ranks with/without bonus offers.

- Three Price levels (from Aug 2010) are available - Price to Stockist (PTS), Price to Retailer (PTR) & Maximum Retail Price (MRP).

- Benchmarked by Government of India, Department of Pharmaceutical – DOP, and NPPA for policy decision on Pricing and monitoring the implementation

- Currently, they have 140 clients of which 66 belong to the Top 100 Pharma companies.

  (Reproduced with permission from AIOCD AWACS).

2. **SMSRC:** Strategic Marketing Solutions and Research Centre (SMSRC), is the leader in the application of prescription research in the Indian and Bangladesh Pharmaceutical Industries.

   Key characteristics of SMSRC are as follows:

   - Strategic Prescription Research (SPR), is the primary service offering.

   - SPR is essentially qualitative Rx Research analysis based on quantitative data that is applicable for Marketing and Corporate strategy formulations in the pharmaceutical industry.

   - SPR's core prescription database constitutes the leading and largest live doctor panel in India and Bangladesh, respectively. SPR's stratified sample database comprises all segments of doctors (+9000), across all tiers of city-town classes (82 centers) in India, and key centers in Bangladesh.

   - SPR is based on live carbon copies of in-clinic prescriptions of doctors; SMSRC collects 100 prescriptions per doctor every two months from its live doctor panel.

- SPR data is available through 'SPRCloud' for clients, via a simple browser-based application, any time - anywhere along with high data security controls.

- SPR showcases doctors' prescription habit analysis over time (bi-monthly and four-monthly basis), to evaluate and tune brand, portfolio, and organization marketing strategy continuously.

- SMSRC is also the leader in Strategic Marketing Consultancy Projects in the Indian & Bangladesh Industry – offering solutions to Brand and Portfolio Management, Doctor Coverage planning, Brand Promotion Prioritization Strategy, Field Force deployment, Doctor Call pattern designs, and other key strategic marketing issues.

(Reproduced with permission from SMSRC)

Let's look at some of the key highlights from SMSRC reports, insights from which can be used in important marketing decision-making like targeting, positioning, etc.

Table 6.1 Top 20 Companies Prescription Share Trend

Company	Prescription Share (%)					
	MAR-JUN 2019	JUL-OCT 2019	NOV-FEB 2020	MAR-JUN 2020	JUL-OCT 2020	NOV-FEB 2021
MANKIND PHARMACEUTICALS	4.8	4.8	4.8	4.8	4.7	4.7
SUN PHARMA	4.2	4.1	4.5	4.3	4.4	4.7
ALKEM LABORATORIES	3.1	3.1	3.1	3.1	3.2	3.3
ABBOTT	3.0	2.9	3.0	2.9	3.1	3.0
CIPLA	2.8	2.7	3.1	3.0	2.8	2.9
DR REDDYS LAB	2.4	2.3	2.5	2.2	2.3	2.4
TORRENT PHARMACEUTICALS	2.1	2.1	2.2	2.1	2.3	2.3
ARISTO PHARMA	2.1	2.1	2.2	2.4	2.2	2.3
GLAXO SMITHKLINE	2.5	2.6	2.6	2.5	2.2	2.2
MACLEODS PHARMACEUTICALS	2.1	2.1	2.3	2.3	2.1	2.1
ZYDUS	2.0	1.9	2.1	2.1	1.9	2.0
IPCA LABORATORIES	1.5	1.6	1.6	1.5	1.7	1.7

Contd...

Company	Prescription Share (%)					
	MAR-JUN 2019	JUL-OCT 2019	NOV-FEB 2020	MAR-JUN 2020	JUL-OCT 2020	NOV-FEB 2021
LUPIN LIMITED	1.5	1.4	1.5	1.5	1.5	1.6
MICRO LABS	1.5	1.5	1.6	1.6	1.6	1.5
INTAS PHARMA	1.3	1.3	1.4	1.4	1.5	1.5
PFIZER	1.3	1.2	1.3	1.3	1.4	1.3
USV LIMITED	1.0	0.9	1.0	1.0	1.2	1.3
ALEMBIC LIMITED	1.1	1.1	1.2	1.3	1.2	1.2
J B CHEMICALS AND PHARMA	1.3	1.2	1.0	1.0	1.1	1.2
FDC LIMITED	1.2	1.3	1.2	1.2	1.2	1.2

Source: SMSRC (Reproduced with permission from SMSRC)

As evident from the above table, based on prescription share, the top 5 companies are Mankind, Sun, Alkem, Abbott, and Cipla. While Mankind followed by Alkem led in the GP/Rural segment, Sun, Abbott & Cipla lead in the specialty segment. Individual companies can carry out their detailed analysis of corporate prescription share across different specialties which can provide valuable inputs for developing their marketing strategies. The detailed analysis of corporate prescription share across different specialties can also provide direction to the new entrant.

The top 20 broad therapeutic categories contribute 81.2 % of the total prescription share as shown in the table below. The contribution of the top 5 broad therapeutic categories is 54.6 % whereas the contribution of the top 10 broad therapeutic categories is 70.5 % respectively to total prescription share.

Table 6.2 Top 20 Broad Therapeutic Categories based on Prescription Share

Therapeutic Category	Rx Share (%)
GI	17.1
PAIN RELIEVERS	11.7
NUTRITIONALS	10.8
ANTIBIOTICS	10.5
CARDIO VASCULAR	4.5
COUGH & COLD	3.7

Contd…

Therapeutic Category	Rx Share (%)
CNS	3.6
DERMATOLOGICALS	3.5
RESPIRATORY	2.9
DIABETICS AND OBESITY	2.2
ANTIALLERGICS	2.0
OPHTHAL/OTOLOGICAL	1.5
STEROIDS	1.5
HORMONES	1.2
PROBIOTICS	1.1
UROLOGICALS	1.1
DENTAL PREPARATION	0.7
HEPATO PROTECTIVE	0.6
THYROID HORMONES	0.5
VACCINES	0.5

Source: SMSRC (Reproduced with permission from SMSRC)

Further analysis of prescription shares of top broad therapeutic categories along with top specialties can be carried out as depicted in the table below.

**Table 6.3 Top 5 Broad Therapeutic Categories along with
Top 5 Specialties Rx Share**

Therapeutic Category	Total Rx Share (%)	Top 5 specialties along with Rx share (%)				
GI	17.1	Gastro (43.3)	Gen Surg (27.1)	Urologist (22.0)	Onco (19.7)	Rural Doc (19.5)
PAIN RELIEVERS	11.7	Ortho (33.2)	Dentist (28.9)	Rural Doc (14.0)	Gen Surg (13.1)	Neuro (12.7)
NUTRITIONALS	10.8	Gynae (21.0)	Nephro (15.6)	Ortho (15.4)	Neuro (14.9)	Onco (12.8)
ANTIBIOTICS	10.5	Dentist (21.3)	Rural Doc (16.0)	ENT (15.1)	Paedia (12.7)	GP (12.6)
CARDIO VASCULAR	4.5	Cardio (25.6)	Nephro (24.4)	Diabeto (17.1)	Cons Phy (8.7)	Neuro (8.7)

Source: SMSRC (Reproduced with permission from SMSRC)

The table above highlights the key specialties that need to be targeted by the companies operating in those respective therapeutic categories. It also provides interesting insights like 'dentists are an important segment to be targeted for pain relievers and antibiotics', 'nephrologists are important target segment for nutritional and cardiovascular products', etc.

Further analysis of prescription shares of top sub-therapeutic categories along with top specialties can be carried out as depicted in the table below.

Table 6.4 Top 5 Therapeutic Categories along with Top 5 Specialties Rx Share

Therapeutic Category	Total Rx Share (%)	Top 5 specialties along with Rx share (%)				
Antiulcer	8.3	Gastro (16.4)	Ortho (12.4)	Gen Surg (12.0)	Dentist (11.5)	Onco (10.9)
NSAID Solids	5.5	Dentist (25.6)	Ortho (13.3)	Gen Surg (7.1)	Rural Doc (6.9)	ENT (6.9)
Cephalosporins Solids	2.6	Rural Doc (4.6)	ENT (4.2)	Dentist (3.6)	GP (3.5)	Gen Surg (3.1)
Antihypertensive	2.6	Nephro (14.2)	Cardio (12.7)	Cons Phy (5.1)	Neuro (4.2)	Chest (4.0)
Analgesic Solids	2.4	GP (3.6)	Rural Doc (3.5)	Cons Phy (2.6)	Chest (2.5)	Cardio (1.7)

Source: SMSRC (Reproduced with permission from SMSRC)

The table above provides important insights for deciding the target audience for those respective sub-therapeutic categories. E.g., nephrologists are the highest prescribers of anti-hypertensive products, dentists are the highest prescribers of NSAID solids, etc.

3. **IQVIA:** IQVIA is a world leader in using data, technology, advanced analytics, and expertise to help customers drive healthcare forward. Key products offered by IQVIA in India are Total Sales Audit, Medical Audit, Subnational Audit, Nepal Audit, and MIDAS. Apart from these syndicated audits, IQVIA also offers customized services to address the specific questions of its clients.

The methodology of data collection is as follows: -

IQVIA has a panel of stockists from which the monthly sell-out is captured and then projected to the universe of stockists (approx.30,000). Post projection they run quality checks through in-house sigma certified QA protocols and finally deliver to clients through business intelligence tools. Data is collected from stockists on monthly basis and reporting is done every month both offline as well as online delivery modes.

Salient Features are as follows:

- IQVIA's sales data capture all India data that can be split into 40 regions. The data can be split into Metro, Class 1, Class 2-4, and rural town classes.

- IQVIA tracks more than 85,000 packs in its database

- Product/pack launch date is captured in the data

    (Reproduced with permission from IQVIA)

MARKETING RESEARCH

Marketing research is the systematic design, collection, analysis, and reporting of data relevant to a specific marketing situation facing an organization.

Companies use marketing research in a wide variety of situations.

1. It can help marketers understand customer satisfaction and purchase behavior.
2. It can help them assess market potential and market share.
3. It can measure the effectiveness of pricing, product, distribution, and promotion activities.

Based on the purpose, there are 3 methods of doing marketing research. These methods are as follows: -

1. Exploratory research - this is done to get an idea about the problem e.g., research done to find out whether doctors would be interested in our new product.

2. Descriptive research - this is done when there is some idea about the problem but done to get more clarity about the characteristics or behavior e.g., what features would the doctor prefer in our new product.

3. Experimental research - this is done to decipher the outcome marketing actions may have e.g., will doctors prescribe more if a new pack is introduced.

Broadly there are two types of marketing research: -

1. Primary research - gathering first-hand information

2. Secondary research - referring to data gathered and published in journals, magazines, reports, websites, etc.

Primary research is mainly of two types: -

1. Quantitative research

2. Qualitative research

1. Qualitative vs Quantitative Research: Quantitative research is all about numbers. It uses mathematical analysis and data to shed light on important statistics about your business and market. This type of data, found via tactics such as multiple-choice questionnaires, can help you gauge interest in your company and its offerings. For example, quantitative research is useful for answering questions such as:

- Is there a market for your products and services?

- How much market awareness is there of your product or service?

- How many doctors are interested in buying your product or service?

- What type of doctors are your best customers?

- What are their prescription habits?

- How are the needs of your target market changing?

- How long are visitors staying on your website, and from which page are they exiting?

Most importantly, because quantitative research is mathematically based, it's statistically valid. This means you can use its findings to make predictions about where your business is headed.

Qualitative research isn't so much about numbers as it is about people – and their opinions about your business. Typically conducted by asking questions either one-on-one or to groups of people, qualitative research can help you define problems and learn about customers' opinions, values, and beliefs. Because qualitative research generally involves smaller sample sizes than quantitative research, it's not meant to be used to predict future performance; rather, it gives you an anecdotal look into your business.

Whereas quantitative research asks short-answer questions that begin with "to what extent," "how much" and "how many," qualitative research asks long-answer questions that begin with "how" and "why." It's particularly useful if you're developing a new product, service, website, or ad campaign and want to get some feedback before you commit a large budget to it. Some typical questions that "qual" research may ask to include:

- Why do you prescribe a particular product?

- Why do you think this product is better than competitive products?

- Why do you prefer prescribing the products of a particular company?

- What should be done to improve this new service to make it more appealing to you?

- What do you think of this new company logo?

- How would you characterize this website design? How friendly and easily navigable is it?

Quantitative research is done by conducting surveys. Qualitative research can be done by conducting an in-depth interview or focus group discussion. Focus group discussion consists of an informal discussion between a small group of 6-10 doctors led by a research expert.

2. Qualitative vs. Quantitative Research

	Qualitative Research	**Quantitative Research**
Objective	To gain a qualitative understanding of the underlying reasons and motivations	To quantify the data and generalize the results from the sample to the population of interest
Sample	A small number of non-representative cases	A large number of representative cases
Data Collection	Unstructured	Structured
Data Analysis	Non-statistical	Statistical
Outcome	Develop an initial understanding	Recommend a final course of action

RESEARCH PROCESS

The research process consists of a series of actions or steps (Fig 6.1) necessary to effectively carry out research and the desired sequencing of these steps.

Formulating the right research objective is extremely important. For example, if you are planning to research to launch a new oral anti-fungal drug, the research objective could be formulated as "To understand the perception of dermatologists about oral antifungal drugs and preference for a particular molecule (if any) along with the reasons".

Once the research objective is clear, other key steps are deciding the respondents, sample size, questionnaire design, data analysis tools, etc.

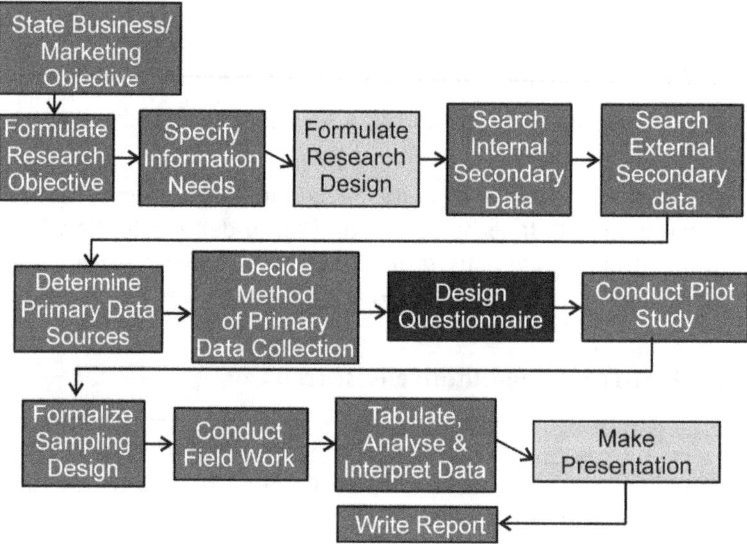

Fig. 6.1 Research Process

SAMPLE QUESTIONNAIRE

1. How many patients do you examine per day?

    (a) < 10   (b)   11-21   (c)   21-31

2. What are the most common ailments your patients suffer from?

    (a) _______   (b)   _______   (c)   _______

3.   What are the medicines you generally prescribe?

(a) _______   (b) _______   (c) _______

4.   When it comes to prescribing a drug / brand, how important are the following parameters and to what extent are you satisfied?

[Based on 5-point scale rank the following,1-Least important/Satisfied,5-Most important/Satisfied]

Parameters	Importance	Satisfaction
Quality		
Price		
Side Effects		
Availability		
Company's Image		
Patient Compliance		
Clinical Data		

5.   How do you rate the following companies on a 5-point scale concerning the following parameters?
(1 = very poor; 2 = poor; 3 = can't say; 4 = good; 5 = very good)

Sr. No.	Parameters	Cipla	Sun	Abbott	Alkem	Mankind
1	Regular Visit					
2	Knowledge					
3	Clinical Data					
4	Quality					
5	Price					
6	Availability					
7	Promotion					

6.   Psychographics

[How strongly do you agree/disagree with the statement?]

(1 = strongly disagree; 2 = disagree; 3 = neutral; 4 = agree; 5 = strongly agree)

Statements	Remarks
I look towards new drugs for patient's health	
I get disappointed with existing drugs	
I am an overworked doctor	
I expect a concise presentation from Medical Representatives (M.R)	
I have a good practice and therefore I am an above-average prescribe of drugs	
I look forward to formal methods of education, seminar, and symposium	
I tend to prescribe generics	
I feel confident to try new drugs	
I believe that the pharmaceutical industry is a good source of information	
I am keen on developing myself as a doctor	
I look at clinical results positively	
I am a successful doctor and therefore do not bother about small companies	
I am keen to medical representatives	
I prescribe only products from multinational companies	
I prescribe products of a company if I am convinced about the quality of products	
I give more importance to the efficacy of drugs than side effects	
I feel that I don't require detailing from MRs	
I give more importance to patient's convenience than pharmacokinetics of drugs	
Samples given by MR's have no impact on prescription	

7. Demographics

   (a) Name:

   (b) Address:

   (c) Qualification:

   (d) Experience

    < 5 years 2) 6-10 years 3) 11-15 years 4) above 15years

SUMMARY

This chapter deals with Marketing Information System & Marketing Research. Key points discussed in this chapter are as follows:

1. Indian pharmaceutical industry is rapidly changing and there is a need to continuously monitor the changes that are taking place in the market, competition, physician prescribing behavior, patient behavior and trade behavior, internal sales trends, etc. It is important to collect, sort, analyze, evaluate and distribute the data for timely decision making. This is done through the Marketing Information System (MIS).

2. Two main pharma databases that play an important role in marketing decision-making are - Retail audit (AIOCD AWACS, IQVIA) & Prescription audit (SMSRC).

3. Marketing research is the systematic design, collection, analysis, and reporting of data relevant to a specific marketing situation facing an organization.

4. Marketing research can help in understanding customer satisfaction and purchase behavior, assessing market potential, measuring the effectiveness of pricing, product, distribution, and promotion activities, etc.

5. Based on the purpose, there are 3 methods of doing marketing research - Exploratory, Descriptive, and Experimental.

6. Marketing research can be Primary research where first-hand data is collected or Secondary research wherein published data is referred.

7. Primary research is mainly of 2 types - Quantitative Qualitative.

8. Quantitative research is all about numbers. It uses mathematical analysis and data to shed light on important statistics about your business and market.

9. Qualitative research can help you define problems and learn about customers' opinions, values, and beliefs.

10. The research process consists of several steps like stating marketing objective, stating research objectives, specifying information need, formulating research design, internal and external data search, deciding method of primary data collection, designing a questionnaire, conducting a pilot study, formalizing sample design, conducting fieldwork, data analysis, presentation, and reporting.

REFERENCES

1. AIOCD AWACS pharmatrack data provided by the company
2. SMSRC prescription audit data,Nov-Feb 2021
3. IQVIA retail audit data provided by the company
4. Nargundkar Rajendra, Marketing Research, Text & Cases, Tata McGraw Hill Education, 2010, pp.20.

CHAPTER **7**

IDENTIFYING MARKET SEGMENTS, TARGETING & POSITIONING

LEARNING OUTCOMES

After reading this chapter, you should be able to:

- Understand the bases for segmenting the pharmaceutical market
- Decide the target market
- Craft the brand positioning

INTRODUCTION

Segmentation is dividing the entire market into different groups(segments) in such a manner that each group is homogenous within (having similar characteristics) but heterogeneous (different characteristics) from one group to another.

Segmenting the market is important for two reasons-

- Each segment may have different needs and hence require different marketing mixes
- Marketing activities can be focused towards a particular segment having maximum potential so that there is optimum utilization of resources

Traditionally, consumer market segmentation has been done in four ways-

1. Geographic - country, region, zone, city, village, etc.

2. Demographic - age, sex, income, education, occupation, social status, etc.

3. Psychographic - attitude, lifestyle, activity, interest, opinion, personality, values, etc.

4. Behavioral - benefits sought, usage, user status, occasion, purchase, etc.

SEGMENTING THE PHARMACEUTICAL MARKET

For a pharmaceutical company, the key customers are healthcare professionals as the revenue generation is through doctor prescription. Having said that the pharma companies are also realizing the significance of communicating with patients and caregivers. According to the Government of India Ministry of Health and Family Welfare Department of Health and Family Welfare, July 2021 report, there are over 12,68,172 doctors in India. These doctors are located in diverse markets and they have distinct needs and wants. No pharma company can connect with all of them. However, the entire doctor population can be divided into smaller groups or segments with distinct needs and wants which can then be selectively contacted based on the potential of each group. This requires a deep understanding of the prescription behavior of doctors and careful strategic thinking. Secondly, with easy access to information and growing health consciousness patients are also actively involved in the choice of the therapy and the brand. Hence, it is equally important to consider patient segments as well.

We will understand this further in this chapter.

According to Prof. Miklos Sarvary,2006 there are broadly two types of segmentation-

- Segmentation based on benefits sought by customers

- Segmentation based on observable characteristics of customers

Let's take an example of a pain killer market. If we consider the case of mild to moderate pain market, it can be segmented based on benefits as those patients requiring 'Quick relief' and those patients for whom 'No side effect' is more important. For a younger patient leading a hectic, active lifestyle the doctor may need a pain killer providing 'Quick relief' whereas in the case of older patients the doctors would more value the benefit of 'No side effect'.

The pharmaceutical market can be broadly divided into two types of segments -

(i)   Doctor segments
(ii) Patient segments

Let's say we are marketing an antibiotic indicated for treating urinary tract infections, the market can be broadly segmented as follows: -

(i)   Doctor segments could be doctors treating urinary tract infections like urologists, nephrologists, physicians, gynecologists, or general practitioners

(ii) Patient segments could be acute cystitis, recurrent cystitis, urethritis, prostatitis, etc.

Patients suffering from the same condition may have varying needs. These may depend upon disease stage, contraindication, compliance, or some other clinically-recognized criteria which are important factors to be taken into account while segmenting the market. Patients may also differ as a result of non-clinical criteria, such as apathy vs. determination (e.g., in the oncology market) or emotional needs (e.g., erectile dysfunction or mental health) and these attitudinal factors can be important in market segmentation (Brian D. Smith, Pharmaceutical Market Europe April 2012).

Based on area of practice, demographic, psychographic, or prescription behavioral parameters, some ways in which doctor segmentation can be done is as follows: -

Area of Practice	**Demographic**
Urban vs rural	Specialist vs general practitioner
North vs South vs West vs East	Male vs Female
Hospital vs medical college vs private practice	Experienced vs young
	Treating rich vs poor patients
	Number of patients
Psychographic	**Prescription behavior with respect to drug/ brand of a particular company**
Attitude towards particular therapeutic category or drug	Awareness of drug/brand-aware/not aware
Preference for drugs marketed by Multinational vs Indian companies	Current prescriber - high, medium, low
Interested in scientific literature vs interested in product samples, sponsorships, promotional items, etc.	Past prescriber
	Never prescribed
Sensitive towards price vs no sensitivity towards the price	Occasion - indications in which prescribe
Technology savvy vs no/minimal use of technology	The benefit sought - efficacy, safety, cost, etc.
Innovator/early adopter vs laggard	Influenced by Key Opinion Leaders (KOLs)/ senior doctors

Fig. 7.1 Market Segmentation

For market segmentation to be effective, all segments must be:

- Distinct - each segment should be different from other segments, thus different marketing mixes will be necessary
- Accessible - buyers can be reached through appropriate promotional mixes and distribution channels
- Measurable - it is easy to identify and measure the segment
- Profitable - the segment should be big enough to provide a stream of constant future revenues and profits
- (Adapted from Kotler, 1984)

TARGETING

Deciding which market segment/s the company should actively pursue in order to generate revenue is called Targeting. Companies choose between adopting undifferentiated, differentiated, or concentrated targeting strategies, a brief description of each is given below.

- Undifferentiated - target all customers with the same marketing mix
- Differentiated - targets several market segments, each with a unique marketing mix
- Concentrated(niche) - selects one segment and concentrates on serving it

How to select a target segment

Selection of target segment can be done as follows: -

Step 1: Identify the 'Market attractiveness' & 'Competitive position' factors for each potential target segment.

Market attractiveness depends on factors like -

- Segment size, growth rate
- Customer needs and behavior
- Macro trends

E.g., if we are looking at the anti-diabetic market, market attractiveness will depend on the size of different sub-therapeutic segments in the anti-diabetic market and their growth rate, patient needs, and prescription behavior of doctors and macro trends like growing incidences of diabetes in India, growing economy, etc.

The competitive position depends on factors like -

- Opportunity for competitive advantage
- Capabilities and resources
- Industry attractiveness

E.g., the competitive position of a particular pharma company would depend on competitive advantage like it is a market leader in a particular segment and enjoys a good image amongst doctors, has research and development capabilities to launch its research molecule and it is operating in a segment where it's difficult for other companies to enter.

Step 2: Once the 'Market attractiveness' & 'Competitive position' factors are identified, assign weights to each factor based on the significance of that factor in that particular market in such a manner that the summation of weights of all factors is equal to "1". Secondly, rate each factor on a scale of "0-10" based on your assessment. Then, multiply each weightage with its corresponding rating to get a total score as indicated in the illustrative example given below.

	WEIGHT	RATING (0-10)	TOTAL
Market attractiveness factors			
Customer needs and behavior	0.5	10	5.0
Segment size and growth rate	0.3	7	2.1
Macro trends	0.2	8	1.6
Total: Market attractiveness	**1.0**		**8.7**
Competitive position factors			
Opportunity for competitive advantage	0.6	7	4.2
Capabilities and resources	0.2	5	1.0
Industry attractiveness	0.2	7	1.4
Total: Competitive position	**1.0**		**6.6**

Fig. 7.2 Assessing Market Segments

Step 3: Plot the total scores for Market attractiveness and Competitive position for each potential target segment in the grid shown below.

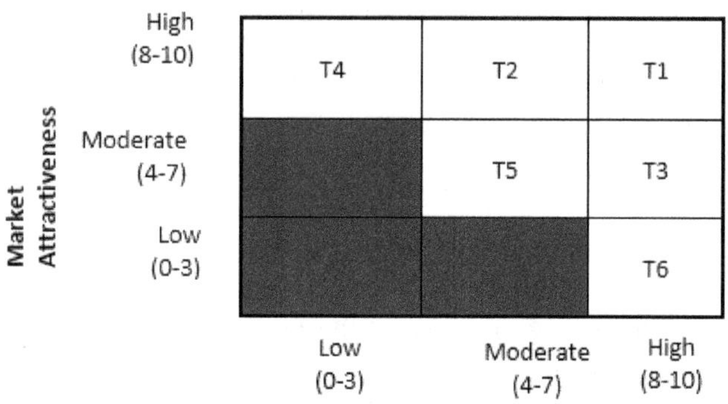

Fig. 7.3 Marketing attractiveness vs Competitive position

T1 is the best segment to target followed by T2 & T3 followed by T4, T5 & T6 segments.

Positioning

Positioning is the act of designing a company's offering and image to occupy a distinctive place in the minds of the target market (Al Ries and Jack Trout, 2000). The result of positioning is the successful creation of a customer-focused value proposition giving a reason why the target market should buy the product (Kotler, Keller, 2013).

Positioning a pharmaceutical brand

Historically, the "Positioning" of a pharmaceutical product was primarily based on the aspects of the product that were most likely to be perceived positively by doctors (Pharmaceutical Executive, Aug 2005). Such an approach worked quite well till there was a constant stream of new drugs. But, over a period of time, new drug discovery has become rare. Secondly, with increasing competition, all brands were talking the same language. Therefore, differentiating one's brand from other competing brands, became essential. This becomes far more critical and a bigger challenge in the case of therapeutic segments where there are hundreds of brands with similar compositions.

How to craft the brand positioning

Steps involved in crafting brand positioning (Kevin Keller,2011) are as follows: -

Step 1: Decide Target Audience (TA)

We have already discussed how to choose the target segment.

Step 2: Define the Frame of Reference (FOR)

Once the target segment is chosen, define the Frame of Reference (FOR) i.e., the therapeutic category to which it belongs. In other words, the products or sets of products your brand is going to compete with. It is important to give consumers a context in which you are trying to establish the superiority of your brand over others in the category. E.g., if you have a brand of Sitagliptin, the FOR could be oral anti-diabetic for the treatment of type 2 diabetes.

Step 3: Identify the Point of Differentiation (POD)

Point of Differentiation (POD) is the attribute or the benefit that your brand offers which is not offered by the competitive brands. Strong brands may have multiple PODs.

The POD should satisfy two important criteria as follows: -

1. Desirability

   There are three key Desirability criteria -

   - Relevance - The target audience must find POD relevant and important
   - Distinctiveness - The target audience must find POD distinctive and superior
   - Believability - A brand must offer a compelling and credible reason for choosing it over other competitive brands.

2. Deliverability

   There are three key Deliverability criteria

   - Feasibility - You should be able to create a POD
   - Communicability - You should be able to communicate in such a manner that your target audience will believe in the brand the POD
   - Sustainability - The POD should be such that it can be reinforced and strengthened over a period of time

Some examples of POD in pharma could be -

- Original research molecule with proven efficacy
- Superior drug delivery system ensuring better efficacy, speed of onset of action lesser side effects, patient compliance, etc.
- Lesser side effects, no drug interaction

Just saying 'better efficacy' or 'better safety' with respect to a particular indication may not help you to differentiate. One need to get deeper into a specific dimension of a particular indication,e.g. does your statin lower LDL or CV prevent outcomes,those are different dimensions(Mike Rea,Oct 2021).

A perceptual mapping technique could be used to identify the POD. Perceptual maps are the visual representations of customer perceptions and preferences. Let's take an example of the analgesic category. The perceptual map based on the 'efficacy' of the analgesic brands and the 'gastric irritation' caused by them may be depicted as shown in fig 7.4. It is evident that brands having higher efficacy (A, B, and E) also cause higher gastric irritation whereas brand C causes low gastric irritation but has low efficacy. There is no brand with higher efficacy and low gastric irritation. From this, it can be concluded that if your brand has moderate to high efficacy and causes low gastric irritation then this can be your POD.

One more technique that can be used to identify POD is by plotting the Opportunity Matrix comprising of 'Importance' & 'Satisfaction' of target customers with respect to key parameters. In this case, the potential target audiences are asked to rate the parameters that are important in the treatment of a particular disease or disorder & how satisfied are they with those parameters. The parameter that is important but having a lower satisfaction score would be the opportunity for your brand to establish the POD.

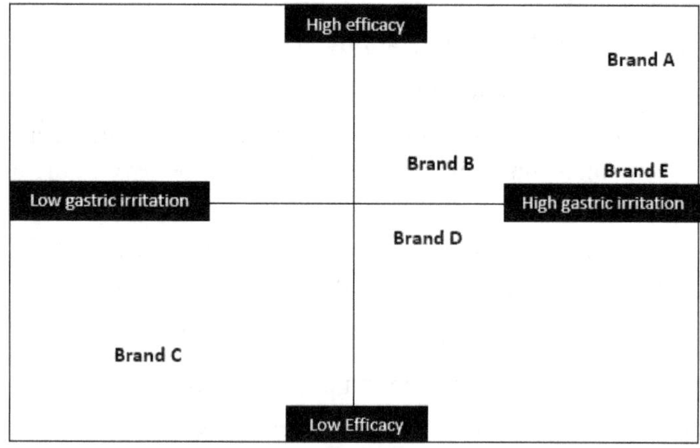

Fig. 7.4 Illustrative example of a perceptual map of the analgesic category

Step 4: Substantiate your POD - give Reason to Believe (RTB)

This is with regards to the believability of the POD. It is important to substantiate the POD (claim) with support, e.g., a clinical trial proving better efficacy compared to other drugs in the category, approved by an authority like Indian Medical Association, Cardiology Society, etc. This is called Reason to Believe (RTB).

Brand Positioning Statement (BPS)

Brand Positioning Statement can be written as follows: -

To ___________(TA) our Brand is ___________(FOR) that is ___________(POD) because ________________(RTB).

E.g., let's say our Brand S is a brand of Sitagliptin, the brand positioning statement can be written as follows-

To Diabetologists, Physicians, and Endocrinologists (TA) treating uncontrolled Type 2 Diabetes patients, our Brand S is oral anti-diabetic (FOR) which provides substantial glucose lowering without compromise enabling physicians to comprehensively treat Type 2 Diabetes early and aggressively (POD) because it significantly reduces HbA1C, FPG & PPG to help get patients to the goal without hypoglycemia and weight gain (RTB).

Brand Positioning Statement is a statement written down in a Brand Plan & it acts as a guiding post to enable consistency across entire marketing activities.

Repositioning

Considering the dynamic nature of the industry, there may be a need to change the positioning of the brand. Repositioning may be necessary due to following reasons (Dave Elzinga,2008):-

- New indication is discovered

- Entry of a new competitor

- Treatment advances that prompt physicians to change the way they think about the disease/condition and/or their treatment approach

- Physician or patient feedback that suggests lack of differentiation

- No clarity about the future direction of a brand

- Conflicting messages leading to confusion among target customers

SUMMARY

This chapter deals with Market Segmentation, Targeting, and Positioning. Key points discussed in this chapter are as follows:

1. Segmentation is dividing the entire market into different groups(segments) in such a manner that each group is homogenous within (having similar characteristics) but heterogeneous (different characteristics) from one group to another.

2. Segmentation can be done based on benefits sought by the customers and based on observable characteristics of customers like demographic, geographic, and psychographic factors.

3. The pharmaceutical market can be broadly divided into two types of segments - Doctor segments & Patient segments.

4. Doctors could be segmented based on the area of practice, demographic, psychographic, or prescription behavioral parameters.

5. Patients could be segmented based on disease stage, contraindication, compliance, or some other clinically-recognized criteria or non-clinical criteria, such as attitudinal factors e.g., apathy vs. determination or emotional needs besides demographic and geographic factors.

6. Selection of target segment can be done based on the assessment of 'Market attractiveness' and 'Competitive position' of each potential target segment.

7. Positioning is the act of designing a company's offering and image to occupy a distinctive place in the minds of the target market.

8. Steps involved in positioning are - deciding the target audience, defining the frame of reference, identifying the point of difference, and giving reason to believe.

9. The point of difference can be identified by applying techniques like perceptual mapping or plotting an opportunity matrix.

10. Some examples of POD in pharma could be original research molecule with proven efficacy, superior drug delivery system ensuring better efficacy, speed of onset of action lesser side effects, better patient compliance, no drug interactions, etc.

REFERENCES

1. Government of India Ministry of Health and Family Welfare Department of Health and Family Welfare, July 2021 Report.

2. Sarvary Miklos, Market Segmentation, Target Market Selection, and Product Positioning, HBS No 501-018, April 17, 2006

3. Smith Brian D., Superior Segmentation: As patients and payers join in the prescribing decision, the industry needs new models of definition, Pharmaceutical Market Europe April 2012.

4. Kotler, P. (1984), Marketing Management, international ed., Upper Saddle River, NJ: Prentice-Hall

5. Urbany Joel E and Davis James H. "Strategic Insight in Three Circles." Harvard Business Review, November 2007

6. Vanderveer Richard B., Position, position, position, Pharmaceutical Executive, Aug 2005

7. Al Ries and Jack Trout, Positioning: The Battle for Your Mind, 2oth Anniversary Edition (New York: McGraw-Hill,2000).

8. Kotler P., Keller K., Koshy A., Jha M., Marketing Management: A South Asian Perspective,14th edition, Dorling Kindersley (India) Pvt. Ltd. 2013.

9. Keller K., Parameswaran M., Jacob I., Strategic Brand Management, 3rd edition, Dorling Kindersley (India) Pvt. Ltd. 2011.

10. Pharmaceutical Positioning book by Mike Rea,Oct 2021

11. Elzinga Dave, Building a Differentiated Brand Positioning, Driving Towards Marketing Excellence, McKinsey & Company, 2008.

CHAPTER 8

PHARMACEUTICAL BRAND AND BRANDING STRATEGIES

LEARNING OUTCOMES

After reading this chapter, you should be able to:

- Understand the concept of product vs brand
- Gain knowledge about pharmaceutical branding
- Understand the concept of brand identity
- Understand the significance and process of developing a good brand name
- Get an insight into branding strategies in pharma

PRODUCT VS BRAND

A product is anything we can offer to a market for attention, acquisition, use or consumption that might satisfy a need or want (Keller,2011).

A brand, according to American Marketing Association (AMA) is a "name, term, sign, symbol, design or a combination of them, intended to identify the goods and services of one seller or group of sellers and to differentiate them from those of competition" (Kotler, Keller 2013). It distinguishes the goods of one producer from those of another. These different components of a brand that identify and differentiate it are called brand elements. The brand however is more than that.

Table 8.1 below depicts the difference between the product and the brand.

Product	Brand
Is tangible	Is intangible
Is built in the factory	Is built in the mind
Can be copied	Is unique
Is an object	Is a personality
Sold by the company	Bought by the customer
Can become outdated	Is timeless

Fig. 8.1 Product vs Brand

Adapted from 60 Minute Brand Strategist: The Essential Brand Book for Marketing Professionals, Idris Mootee, 2013, John Wiley & Sons, Inc., Hoboken, New Jersey

Five levels for a product (Theodore Levitt,1980) can be defined as follows:

1. The ***core benefit level*** is the fundamental need or wants that consumers satisfy by consuming the product or service.

2. The ***generic product level*** is a basic version of the product containing only those attributes or characteristics necessary for its functioning but with no distinguishing features. This is a stripped-down, no-frills version of the product that adequately performs the product function.

3. The ***expected product level*** is a set of attributes or characteristics that buyers normally expect and agree to when they purchase a product.

4. The ***augmented product level*** includes additional product attributes, benefits, or related services that distinguish the product from competitors.

5. The ***potential product level*** includes all the augmentations and transformations that a product might ultimately undergo in the future.

Example: Analgesic	
1. Core Benefit	Relieving pain
2. Generic Product	Analgesic tablet providing pain relief
3. Expected Product	An analgesic tablet that relieves pain without causing any side-effect
4. Augmented Product	An analgesic tablet that is formulated in such a manner that it provides the benefits of quick relief, the action lasts for a longer duration, avoids side effects, and is convenient for the patient to administer
5. Potential Product	The analgesic formulation releases drug only at the site of pain

Fig. 8.2 Example of Different Product Levels

PHARMACEUTICAL BRANDING

Pharmaceutical branding originated more than 100 years ago when Thomas Beecham recognized the importance of branding his safe and effective new laxative 'Beechams Pills'. This started a new trend in the marketing of medicines by attaching a personal guarantee of the product's effectiveness, enabling it to stand out from the plethora of other products on the pharmacist's shelves which were likely to cause as much harm as good (Tom Blackett and Rebecca Robins, 2001).

Historically, pharmaceutical companies focused more on marketing their products than marketing through brands. They focused too heavily on product attributes and neglected other important facets of the branding paradigm, therefore managing products and not brands (Moss,2001). However, with the slowing down of new drug discovery resulting in the slowing of industry growth, firms have been searching for new ways to maintain their brand positions (Schuiling and Moss, 2004). Consequently, pharmaceutical companies have had to embrace marketing and branding strategies to a greater extent than they have been in the past (Blackett, 2005).

Moreover, branding offers several benefits as follows: -

- A brand can move beyond functional benefits by providing additional tangible (rational) and intangible (emotional) values (Blackett and Harrison, 2001).

- Branding helps to create differentiation that is difficult for rivals to copy, this is a source of competitive advantage (MacLennan, 2004).

- Brands reduce perceived risk, build trust, and stimulate excitement that results in preference, long-term commitment, and prevents substitution (Kapferer, 2012).

- Brand assets include awareness, image, reputation, and perceived brand personalities. These assets will produce brand strength (market share, market leadership, loyalty, and price premium) that will ultimately lead to brand value (the ability for brands to deliver profits) (Kapferer, 2012).

Over the years the pharmaceutical industry has undergone several changes. According to recent estimates, the cost to develop a new molecule could be as high as $2.8 billion (Olivier J. Wouters, Jan 2020). Therefore, rather than focusing on launching new molecules, there is more emphasis on brand building. There is widespread acceptance of brands as strategic assets. Since the Covid-19 pandemic, there has been rapid adoption of digital channels by doctors as well as patients. Patients today are acquiring much more information from various sources available and are actively engaging with their physicians and caretakers. Consumers are online and on social media, and that's where pharmaceutical companies also need to be. Thus, it is no longer enough to talk only about product attributes. A pharmaceutical brand needs to tell compelling stories based on the needs and expectations of physicians as well as patients.

BRAND IDENTITY

Brand identity comprises visible elements of a brand like a logo, color, design, etc. that identify and differentiate the brand in the consumer's mind. Besides names, logos, labels, distinct colors, fonts, taglines, packaging, etc. shapes, objects, sounds, smells, etc. can be developed as brand identities. Since a brand identity is tangible, it acts as a distinguishing feature in the minds of consumers and gives them a perception about the business it denotes or is associated with (Kevin Budelmann 2010).

According to Alina Wheeler (2013), Brand identity is tangible and appeals to the senses. You can see it, touch it, hold it, hear it, watch it move. Brand identity fuels recognition amplifies differentiation and makes big ideas and meaning accessible. Brand identity takes disparate elements and unifies them into whole systems.

Brands too have personalities just like human beings. Brand personality can be conceptualized as the symbolic meaning a brand acquires (Sung and Kim, 2010) and formally defined as "the set of human personality traits that are both applicable to and relevant for brands" (Azoulay and Kapferer, 2003, p. 151). Research conducted by Kapferer (1998, 2012) concludes that generalist doctors and specialists can attribute human personality traits to medicines. Using a scale of 15 personality traits such as caring, empathetic, and creative, he established a correlation between prescription levels and certain personality traits.

Another research conducted by Leonard and Katsanis (2013) found that consumers do attribute human personality traits to prescription drugs and that prescription drug brand personality as perceived by consumers has two distinct dimensions: Competence and Innovativeness as depicted in fig below: -

Competence	Innovativeness
Dependable	Unique
Reliable	Innovative
Responsible	Original
Successful	
Stable	
Practical	
Solution-Oriented	

Fig. 8.3 Prescription Drug Brand Personality

SIGNIFICANCE AND PROCESS OF DEVELOPING GOOD BRAND NAME

A 'brand name' is the most important of all brand elements. It is the face of the brand, the first indicator of the brand for all awareness and communication efforts. It creates an immediate and usually lasting impression that will shape the perception of the brand (Gronlund Jay,2013).

Developing a good brand name is extremely important in pharmaceutical marketing. A doctor is bombarded with hundreds of brands with similar messaging on daily basis and it is very difficult for a doctor to remember all those names. Moreover, a doctor has to recall the brand name at the spur of the moment while writing a prescription. Thus, the significance of a good name that is easy to remember and recall. So, what are the characteristics of a good brand name? Let's have a look at the process of creating a good brand name.

How to create a good brand name?

Following are the important aspects of a good brand name.

- Must be simple and memorable
- Should be easy to pronounce and sound good
- Should be easy to write (prescription)
- Should be easy to associate with the disease, therapeutic class, or patient condition
- Should reflect brand image and personality
- Should be distinct from competitors
- Should be protectable (trademark patent)
- Should be such that can be carried internationally

"Phonologics" is often used for creating new brand names. Research by brand consultants and linguists indicates that there is a subconscious relationship between certain speech sounds and emotions, e.g., the use of "x", "z" and "c" all imply power and innovation (Gronlund Jay,2013). Some examples of pharmaceutical brand names leveraging this are Nexium, Celebrex, Zyban, Zithromax, etc.

Given below are some more examples of pharmaceutical brand names using a combination of word parts and individual letters reinforcing the desired brand identity (Gronlund Jay,2013).

- Prozac: The letter "p" implies daring and activity and supports its "ac" ending, which suggests action. The letter "z" also implies action or speed of recovery.
- Zoloft: "zo" means "life" in Greek and "loft" suggests elevation. The beginning "z" and ending "ft" are very attention-getting. The letter "z" is also the highest-rated for fast and active, yet comfortable.

Trademark is a legal synonym for brand name or brand identity. After a trademark is selected, it is essential to know whether it is available (not in use by others). For this purpose, it is important to check in relevant trade magazines, directories, online mediums, and trademark journals published by Trademark Offices, to find out if there is any other party using the trademark or any other similar brand identity (AditiThakurVerma,2012). There are three types of brand names as follows: -

1. Descriptive

2. Suggestive or Associative

3. Abstract

Descriptive names describe the product and are therefore easy to understand. Examples of descriptive names are -

- Zincovit (combination of zinc and vitamins)

- Shelcal (Calcium derived from the shell)

- Easibreathe (which eases breathing)

Suggestive or Associative names partially describe what the product is. They communicate some benefits or features of the product. Examples of suggestive names are-

- Asthalin (used to relieve symptoms of asthma)

- Sinarest (relieves sinus congestion in the common cold)

- Flexijoints (for making joints flexible)

Abstract names have no inherent meaning and nothing is described. Examples of abstract names are-

- Zedex

- Zinetac

- Dexorange

BRANDING STRATEGIES IN PHARMA

Branding strategies also referred to as brand architecture, reflect the number and nature of both common and distinctive brand elements (Kotler, Keller,2013). Some of the branding strategies adopted by pharmaceutical companies are described below.

A. **Product Branding Strategy:** The brand is promoted exclusively so that it acquires its own identity and image. It does not share other products and does not take on company associations. Most pharmaceutical companies follow this strategy.

B. **Corporate Branding or Umbrella Branding Strategy:** In this case, the brand names are linked to the company brand name. E.g., Mankind pharma follows this strategy for certain brands like Amlokind, Moxikind, Cefakind Caldikind, Nurokind, Dolokind, Ferikind, Lipikind, Lipirose, Losakind, Telmikind, Pantakind, Rabekind, etc.

C. **Drug/Therapeutic Class Branding:** The name is partly based on the therapeutic class to which the product belongs.

E.g.Ciplox(ciprofloxacin), Norflox(norfloxacin), Oflox(ofloxacin) - all are quinolone antibiotics marketed by Cipla

D. Condition Branding or Disease Branding: Condition branding or Disease branding means looking beyond product-centric marketing to the broader factors surrounding the treatment. Product branding tells consumers about a solution but not about the problem which the solution addresses. **Condition** branding educates consumers, physicians, and other stakeholders about the problem. When the **condition** and product branding are well-coordinated, each enhances the effectiveness of the other, raising patient health and brand sales (**Angelmar Reinhard** et.al. 2007). This could be done in conditions like AIDS, cancer, schizophrenia, etc.

For example, MSD pharma launched the brand, Gardasil which is a Human Papillomavirus (HPV) vaccine to prevent HPV-related cancers.

Disease branding would be appropriate if following criteria (Vince Parry, Oct 2007) are fulfilled:-

- Does the product impact a disease or condition in a new way?

- Are there stigmas/social concerns about the condition that can hinder customer self-identification and the patient-physician dialogue?

- Does your product have significant benefits for a little-known condition?

E. Brand Extension: The brand extension involves using an existing brand name to launch a new product to target different market segments within the same category(Line Extension) or to enter a different product category(Category Extension).

Examples are -

- GSK extended its brand Crocin(paracetamol 500 mg) to Crocin pain relief(paracetamol 650 mg & caffeine 50 mg) & Crocin Cold & Flu (paracetamol 500 mg + caffeine 32 mg + phenylephrine 10 mg).

- Cipla extended its brand Novamox (amoxycillin) to Novamox-LB (amoxycillin + lactic acid bacillus spores) & Novamox-AX (amoxycillin + ambroxol)

F. Co-operation: These include Joint promotion, Alliance, Cobranding, Joint venture, Licensing, Franchising, etc.

Examples are-

- Novo Nordisk manufactures and distributes insulins in partnership with Abbott and Torrent Pharma. Torrent manufactures and distributes human insulin of Novo Nordisk. Abbott distributes Novo Nordisk human insulin under the top-selling brand Human Mixtard.

- Merck has licensed its anti-diabetic brands Januvia and Janumet to Sun Pharma for increased market penetration and access to these drugs for patients in India.

SUMMARY

This chapter deals with product vs brand, brand identity, pharmaceutical brand naming, and branding strategies. Key points discussed in this chapter are as follows:

1. A product is tangible while the brand is intangible and is built in the customer's mind.

2. Historically, pharmaceutical marketing has been product-centric marketing. However, with the slowing down of new drug discovery resulting in the slowing of industry growth, firms have been searching for new ways to maintain their brand positions.

3. Branding offers several benefits like creating differentiation that is difficult for rivals to copy, reducing perceived risk, building trust, and stimulating excitement that results in preference, long-term commitment, and prevents substitution.

4. Brand assets include awareness, image, reputation, and perceived brand personalities. These assets will produce brand strength (market share, market leadership, loyalty, and price premium) that will ultimately lead to brand value (the ability for brands to deliver profits).

5. Brand identity comprises visible elements of a brand like a logo, color, design, etc. that identify and differentiate the brand in the consumer's mind. Besides names, logos, labels, distinct colors, fonts, taglines, packaging, etc. shapes, objects, sounds, smells, etc. can be developed as brand identities.

6. Developing a good brand name is extremely important in pharmaceutical marketing. A doctor is bombarded with hundreds of brands with similar messaging on daily basis and it is very difficult

for a doctor to remember all those names. Moreover, a doctor has to recall the brand name at the spur of the moment while writing a prescription. Thus, the significance of a good name that is easy to remember and recall.

7. A brand name should be simple, memorable, easy to write, easy to pronounce, sound good, should be easy to associate with the disease, therapeutic class, or patient condition, should reflect brand image and personality, should be distinct from competitors, should be protectable (trademark patent), should be such that can be carried internationally

8. "Phonologics" is often used for creating new brand names. Research by brand consultants and linguists indicates that there is a subconscious relationship between certain speech sounds and emotions, e.g., the use of "x", "z" and "c" all imply power and innovation.

9. Brand names can be broadly classified into 3 types - descriptive, suggestive/associative and abstract

10. Different types of branding strategies adopted by pharmaceutical companies are product branding, corporate branding, drug/therapeutic class branding, condition branding, brand extension, brand cooperation, etc.

REFERENCES

1. Keller K., Parameswaran M., Jacob I., Strategic Brand Management, 3rd edition, Dorling Kindersley (India) Pvt. Ltd. 2011.

2. Kotler P., Keller K., Koshy A., Jha M., Marketing Management: A South Asian Perspective, 14th edition, Dorling Kindersley (India) Pvt. Ltd. 2013.

3. Moote Idris, 60 Minute Brand Strategist: The Essential Brand Book for Marketing Professionals, 2013, John Wiley & Sons, Inc., Hoboken, New Jersey.

4. Theodore Levitt "Marketing Success Through Differentiation - of Anything", Harvard Business Review (Jan-Feb 1980):83-91.

5. Blackett Tom and Robins Rebecca, Brand Medicine: The Role of Branding in the Pharmaceutical Industry, Interbrand, 2001.

6. Moss, G. (2001), "Pharmaceutical brands: do they really exist?", International Journal of Medical Marketing, Vol. 2 No. 1, pp. 23-32.

7. Schuiling I, Moss G (2004). How different are branding strategies in the pharmaceutical industry and the fast-moving consumer goods sector? J. Brand Manage. 11(5):366-380.

8. Blackett K, Tom F (2005). Branding in the pharmaceutical industry packing packaging

9. Wouters O J, McKee M, Luyten J. Estimated Research and Development Investment Needed to Bring a New Medicine to Market, 2009-2018. JAMA.2020;323(9):844–853.

10. Blackett, T. and Harrison, T. (2001), "Brand medicine: use and future potential of branding in pharmaceutical markets", International Journal of Medical Marketing, Vol. 2 No. 1, pp. 33-49.

11. MacLennan, J. (2004), Marketing Planning for the Pharmaceutical Industry, 2nd ed, Gower Publishing, Ashgate.

12. Kapferer, J-N. (2012), The New Strategic Brand Management: Advanced Insights & Strategic Thinking, 5th ed., Kogan Page, London.

13. Budelmann L, Kim Y and Woznaik C, Brand Identity Essentials: 100 Principles for Designing Logos and Building Brands, Rockport Publishers, USA, 2010.

14. Wheeler Alina, Designing Brand Identity: an essential guide for the whole branding team, 2013, Published by John Wiley & Sons, Inc., Hoboken, New Jersey.

15. Sung, Y. and Kim, J. (2010), "Effects of brand personality on brand trust and brand affect", Psychology & Marketing, Vol. 27, No. 7, pp. 639-661.

16. Azoulay, A. and Kapferer, J-N. (2003), "Do brand personality scales really measure brand personality?", Journal of Brand Management, Vol. 11 No. 2, pp. 143-155.

17. Kapferer, J-N. (1998), The Role of Branding in Medical Prescription, Research Paper Series, HEC Graduate School of Management, Paris.

18. Leonard Erica and Katsanis Prevel Lea, The dimensions of prescription drug brand personality as identified by consumers, Journal of Consumer Marketing, Volume 30 Number 7, 2013, pp 583–596.

19. Gronlund Jay, Basics of Branding: A Practical Guide for Managers, Business Expert Press, 2013.

20. Thakur Aditi Verma, Branding and Business Management: Leveraging Brand Names for Business Advantage, Journal of Intellectual property Rights, Vol.17, Sept 2012, pp.374-384.

21. Reinhard Angelmar, Sarah Angelmar, Liz Kane, Building Strong Condition Brands, Journal of Medical Marketing, Sep 2007, Vol. 7 Issue 4, p341-351

22. Parry Vince, Disease Branding: What Is It, Why It Works, And How To Do It, Guide To Branding, October 2007, A Supplement To Pharmaceutical Executive.

CHAPTER 9

NEW PRODUCTS – GROWTH DRIVERS IN PHARMACEUTICAL INDUSTRY

LEARNING OUTCOMES

After reading this chapter, you should be able to:

- Understand the concept of new product
- Understand the objectives of launching new products
- Understand the new product development process
- Understand the pharmaceutical product life cycle
- Get an insight into product strategies for growth

NEW PRODUCT

New products could be classified into the following six categories (adapted from Booz, Allen, and Hamilton, 1982) as follows: -

1. New to the world - these are new drug discoveries that create an entirely new market e.g., AIDS vaccine, an anti-cancer drug, etc.

2. Products new to the company - these are products new to the company but not new to the market. They allow a company to enter an established market for the first time.

3. Additions to existing product lines - these are products that supplement established lines, e.g., new dosage forms or new variants of flavor, package, etc.

4. Improvements and revisions of existing products - these products replace existing products and offer better efficacy, safety, patient compliance, etc.

5. Repositioning - here existing products are targeted to new markets or market segments

6. Cost reductions - these are products that provide similar performance at a lower cost

The new product could be launched by a company in two ways-

- Developing a new product on its own (focusing on research and development)

- Acquiring another company, another firm's patent, or buying a license or franchise

Very few companies in India focus on investing in research & development to discover new drugs. The majority of new products launched by pharma companies in India are incremental improvements over existing products or by acquiring other firms' patents or buying a license or franchise.

OBJECTIVES OF LAUNCHING NEW PRODUCTS

Pharmaceutical companies have long relied on new products for their growth. The majority of Indian companies have grown over the years by launching several new products which are "me-too" in nature. Big multinational companies have grown by launching their original research molecules in India.

Secondly, existing products over a period of time show a decline in sales or get stagnated due to various reasons like increased competition, newer better therapies are launched, changing doctor or patient preferences, etc. Hence, a pharmaceutical company needs to stay tuned to the changes taking place in the market, closely track the sales and customer perception of their existing products and launch new products which are improvements over existing products.

Although new product launches primarily accounted for the major growth in the past, the Indian pharma industry is currently witnessing a decline in new product launches. New products account for 24 percent of the pharma growth in FY20, compared to 45 percent in FY17(Business Standard, May 11, 2020). Companies are focusing more on building brands and are launching brand extensions to premiumise their product portfolio. Factors that are adversely impacting new product launches are -

- Multiple launches increase marketing costs and lead to a cluttered portfolio.

- Rapid growth in the trade generics market as consumers are preferring low-priced medications

Covid-19 pandemic has once again given a boost to new product launches primarily in the anti-infectives, nutritionals, respiratory and analgesic categories besides cardiac, anti-diabetic, gastro, and derma categories. According to AWACS PharmaTrac MAT OCT 2021 data, 4446 brands have been launched in the last 24 months contributing INR 5685 crores to MAT OCT 2021 value.

New Product Development Process

New product development is one of the key elements of a pharmaceutical company in terms of its growth and development. In rapidly changing business conditions, the success of new product development processes highly depends on the technological knowledge and the company's ability to transform it into value (Ecem, Meşepınar et.al.,2021). Nerkar and Roberts (2004) stated that in addition to these factors, manufacturing, marketing, sales, and distribution of those products should be managed strategically. According to the study conducted by Yousefi et al. (2017), critical success factors of new product development are managerial capabilities and management commitment to new product development projects.

New product development could be the development of a new drug (innovator drug) or a generic product (product launched by another company after the patent expiry of the innovator drug company). Let's first look at new drug development.

New Drug Development

New drugs take an average of $2 billion and more than ten years to reach the market after discovery. (Anna Heid and Ivan Ostojic, Oct 2020).

Out of the 39 blockbuster launches (those with more than $1 billion in forecasted sales) due to take place between 2021 and 2025, 22(i.e., 56 percent) are expected to come from first-time launchers. That would represent a sizeable increase over the period from 2016 to 2020, when less than 20 percent of blockbusters came or were expected to come, from first-time launchers (Eylul Harputlugil, et.al., Feb 2021)

New drug development is a complex process. It comprises the following key stages (adapted from Amol Deore et.al. 2019 and Yeolan Lee et.al. 2019): -

1. **Target identification and validation** - The firm conducts research to identify a drug target (molecules associated with particular diseases that can potentially produce desirable therapeutic effects).

2. **Lead identification and optimization** - A lead is a synthetically stable, feasible, and drug-like molecule active in primary and secondary assays with acceptable specificity, affinity, and selectivity for the target receptor.

3. **Product characterization** - When any new drug molecule shows a promising therapeutic activity, then the molecule is characterized by its size, shape, strength, weakness, use, toxicity, and biological activity.

4. **Formulation and development** - The pharmaceutical formulation is a stage of drug development during which the physicochemical properties of active pharmaceutical ingredients (APIs) are characterized to produce a bioavailable, stable, and optimal dosage form for a specific administration route.

5. **Preclinical research** - Preclinical research in the drug development process involves the evaluation of drug's safety and efficacy in animal species that conclude to the prospective human outcome. After animal testing results show positive signs and the commercial potential of a lead, the lead becomes a drug candidate.

6. **Investigational New Drug filing** - The firm files an investigational new drug (IND) application to the Food and Drug Administration (FDA) and initiates three phases of clinical trials.

7. **Clinical Trials** - The firm tests the efficacy and toxicity of the drug candidate through these trials:

   Phase I includes 20 to 100 individuals with or without the disease,

   Phase II includes several hundred patients with the disease, and

   Phase III includes 300 to 3,000 patients that have the disease.

8. **New Drug Application** - After the firm collects sufficient evidence that shows the efficacy and safety of its drug candidate, it files a new drug application (NDA) for FDA review to obtain market approval.

9. **Approval** - The firm can sell the drug in the market after approval.

Selecting therapeutic areas or indications to invest in is driven by 'medical need' and the prevalence of the disease (Tamimi and Ellis, 2009). Additional factors also include technical feasibility, research, and development costs, and commercial considerations such as competition in the marketplace and potential market share.

NEW GENERIC PRODUCT DEVELOPMENT PROCESS

As stated earlier, the majority of new products launched in India are generic products. Generic drug entry could be in 3 ways (Vikram Pratap Sigh Thakur, 2012):

- Safe entry i.e., at the end of exclusivity and/or patent expiry

- At-risk entry, by trying to challenge the validity of the patent

- By exploring licensing and compulsory licensing options

Different stages in new product development (adapted from Ecem, Meşepınar et.al., 2021, Rizwan Raheem Ahmed et.al. 2014 and Atanu Chaudhuri,2013) are as follows: -

1. **Idea generation - identification of a product:** The first and foremost step in new product development is the identification of molecule; i.e., the active ingredient that a company wishes to launch. New product ideas can come from anyone within the company or customer or channel partner.

2. **Screening of products:** The next important stage is deciding which product(molecule) to go ahead with. Several factors are taken into account while screening a molecule to find its merits and demerits. Some of the important factors are as follows:

   - **Company's objectives** - what is the aim:
     o Whether to enter a new product category
     o To strengthen an existing product line
     o To maximize market share
     o To maximize revenue
     o To maximize profit

   - **Patent status** - consider the following factors:
     o Review patent period
     o Is the company holding patent active in India
     o Has the company holding a patent taken legal action against any other company?
     o Validity period left for the patent

- **Market potential**
  o Market size
  o Growth rate
  o Prevalence of disease for the treatment of which the product is going to be used
- **Special requirements** - consider the following factors:
  o Are there any special manufacturing/ distribution requirements like humidity/temperature/moisture, storage conditions, packaging material, etc?
  o Does the company have such a facility? What will be the cost in case the company does not have the facility?
  o Can the current sales force handle the new product? Is special training required?
- **Competition**
  o Number of brands of the same molecule in the market
  o Number of players
  o Acceptance of the molecule by doctors

3. **Sourcing of raw material:** A pharmaceutical product has to conform to standard specifications all the time. Every drug has its standard characteristics like dissolution time, shelf-life physical appearance and stability in form, etc. This stage involves -

- Raw material identification
- Vendor identification and indenting
- Procurement of raw material for lab development

4. **Research & design:** This stage involves-

- Sample development
- Analytical method development
- Process optimization
- Finalizing analytical method
- Preparation of development report

5. **Pilot batch manufacturing & conducting stability studies:** The trial batch is manufactured while maintaining real-life conditions. Once the batch is manufactured, it is kept under different conditions

like room temperature, high temperature, high humidity, sunlight, etc. for 3-6 months and a test is performed each month to see the stability of a product.

6. **Drug Master File (DMF):** Drug Master File (DMF) containing comprehensive, accurate, and precise information about the product is prepared and filed along with Marketing Authorization Application (MAA)with Central Drug Standard Control Organization (CDSCO).

7. **Marketing Strategy Development:** Parallel to trial batch testing, the company starts working on its marketing plan. A typical marketing plan covers the following areas.

- **Market overview:** Size of the segment, profile of the segment, growth rates, prospects

- **SWOT analysis:** Company's strength and weaknesses, environmental threats and opportunities

- **Competition analysis:** Number of players, market shares, growth rates, SWOT

- **Product profile:** Classification, molecule structure, mode of action, advantages/disadvantages for the segment, dosage form, dosage regimen

- **Clinical profile:** Indications, efficacy, side effects, precautions, other pharmacological aspects, clinical reports, etc.

- **Product positioning:** Based on the uniqueness, point-of-differentiation, image of the brand that the company wants to create in the customer's mind

- **Marketing/Sales objectives:** Qualitative and quantitative objectives.

- **Promotional strategy:** Message(s) to be communicated, material to be used, sampling to be done, other activities like clinical trials, user studies, seminars, symposia, advertisements, etc.

- **Sales & Distribution strategy:** Areas/segments to be covered, doctors to be visited, indications/diseases to be focused according to different target segments, etc.

- **Financials:** Detailed revenue and expenses statement

8. **Test Marketing/clinical Trial/user Trial:** Depending upon the nature of the product, a company may decide to go for a clinical trial or user study to enhance the confidence of the medical

profession in the efficacy and/or safety of the product. It can be a fully protocol-controlled trial or just a user study where the doctor can judge the results obtained through the product by using the free samples provided to him/her. The results, if come out to be positive, become a very important promotional tool for the company. The trial can be either pre-launch or post-launch depending upon the situation and nature of the product.

9. **Commercialization:** Commercialization starts with sending the stock to distributors. The sales team plans the promotional activities given the availability of the product and starts visiting the doctors, in whose vicinity the product has reached the chemist shelves.

NEW PRODUCT LAUNCH SUCCESS

Pharma companies need to think of innovative strategies to succeed with the new product launches. Replicating successful launch strategies from the past is no longer a safe option since radical changes in physicians' preferences and behaviors have taken place due to the Covid-19 pandemic. Companies need to work on enabling the sales force to engage with doctors in innovative ways, analytics-enabled engagement with healthcare professionals, adopting innovative patient channels and services, providing personalized content digitally, adopting omnichannel distribution strategy, and exhibiting operational excellence.

Given below are the new product launch tracking guidelines (Table 9.1).

Table 9.1 New Product Launch Tracking

Activity	Due Date	Bottle-necks	Suggested Action Plan
Product Development Activities	***Responsibility***		
Order for material			
Development Work			
Packaging Development:			
- Stability Studies			
- Costing Data			
- PA documentation & Submission			
- Artwork preparation			
- Foil/Carton/Leaflet/Labels/Catch cover			

Contd...

Activity	Due Date	Bottle-necks	Suggested Action Plan
- Placing orders for RM & Packaging Material			
FDA Submission / Approval			
Medical Activities	*Responsibility*		
Medical Rationale			
Statement of Claims			
Approximate time for			
- Clinical Trials			
- Bio-equivalence Study, if required			
- Bio-availability			
- DCGI approval			
Marketing Activities	*Responsibility*		
Brand Name & Logo			
Proposed Dosage form			
Brief Description (Choice of color, flavor, shape, etc.)			
Unit Pack			
Sales Requirement (Month-wise for 1 year)			
Launch Requirements			
- Sales			
- Samples			
Letter of consent (L.O.C.)			
Proposed Price			
Promotional Material			

Activity	Due Date	Bottle-necks	Suggested Action Plan
Materials Activities		*Responsibility*	
Placing a purchase order for			
- Active raw material			
- Excipients			
Packaging Material - Delivery			
- Aluminum foil			
- PVC Film			
- Cartons			

Contd...

Activity	Due Date	Bottle-necks	Suggested Action Plan
- Shippers			
- Leaflets			
- Catch covers			
- Labels			
Manufacturing Activities		*Responsibility*	
Plant & Machinery			
Any changes required			
Special Requirements			
- Light			
- Humidity			
- Temperature, etc.			
Safety Requirements			
Plan for 1st Production Batch			
Validation Protocol			
Packaging of 1st Production Batch			
Dispatch of 1st Production Batch			

PHARMACEUTICAL PRODUCT LIFE CYCLE

As newer drug molecules/therapies are launched in the market they replace the older molecules/therapies. Thus, a pharmaceutical product passes through four stages of introduction, growth, maturity, and decline (adapted from Kotler, Keller, 2013) as follows.

1. **Introduction** - launch phase where sales are low and there are no profits

2. **Growth** - period of rapid acceptance by doctors resulting in growing sales revenue and improvement in profits

3. **Maturity** - a slowdown in sales growth as the product has achieved acceptance of most potential prescribers. Profits stabilize or decline due to increased competition

4. **Decline** - sales on declining trend and profits are eroding

Pharmaceutical companies adopt various strategies to effectively manage the product life cycle to stay ahead as stated in Table 9.2 below.

gies/ ges	Introduction	Growth	Maturity	Decline
ct	Depending on the therapeutic category, the company may launch one or more dosage form/s in the most important strengths.	Extend Product Depth with new dosage forms/strengths/features like flavors or packaging.	Maintain the complete range and try to focus on the most profitable one to gain maximum advantage. Also, identify and explore new indications /segments	Continue as long as the product is beneficial commercially and then withdraw.
	Depends on whether the product comes under DPCO or not. In the case of DPCO, the price will be as per the regulation. In case the product is out of DPCO, then the company may adopt penetration pricing (in case of the me-too product) or skimming pricing (in case of the innovative product), or competitive pricing depending on the market and the competition.	Same as fixed at the time of launch	Same as fixed at launch. Some companies may choose to reduce prices to gain/maintain market share depending on the competition.	Companies may choose to reduce prices to counter the erosion of market-share
	Selective, available at leading pharmacy outlets only.	Extend the availability of the product at more pharmacy outlets in the	Available at all pharmacy outlets from where there is the	Selective availability in pharmacy outlets, phase out those which are not

tegies/ ages	Introduction	Growth	Maturity	Decline
otion	Heavy sampling for trial generation, focus on opinion leaders/consultants, elaborate leave behind literature/brochures, user/clinical trials, seminars/synopsis/panel discussions, advertising in medical journals, valuable brand recall items, conference sponsorships, and digital marketing. Direct mailers and teasers may also be used at this stage.	Moderate sampling, focus on coverage of doctors and call-rate, focus on consultants as well as general physicians, moderate advertising in medical journals, reminder brochures, reminder brand recall items, conference sponsorships, digital marketing engaging the doctors.	Stress on brand's competitive advantages and encourage taking prescription share from competitors. Continue to focus on loyal prescribers. Sampling mostly on demand, reminder brochures, and reminder brand recall items.	Selective and focused promotion to retain loyal prescribers.

PRODUCT MIX

The product mix of a company consists of various products and SKUs (stock-keeping units) of each product. There are 3 main dimensions of product mix as follows:

- Product width - this refers to the number of product lines that the company carries, e.g., many companies in India have multiple product lines like anti-infectives, painkillers, antacids, anti-ulcerants, anti-diabetics, anti-hypertensives, multivitamins, etc.

- Product length - this refers to the number of items(products) the companies carry in a particular line, e.g., companies like Cipla, Alkem, have a wide range in their anti-infectives product line

- Product depth - number of variants each product offers, e.g., different strengths of the drug, different pack sizes, different flavors, etc.

As a growth strategy, the company keeps on adding new products and SKUs to its product mix. However, after a certain period, the product mix becomes too big to handle. This gives rise to the need for product rationalization. Therefore, a company needs to monitor the performance of each product regularly to add or prune the product mix to stay profitable.

PRODUCT STRATEGIES FOR GROWTH

Ansoff's matrix gives the following four product strategies for growth.

1. **Market penetration:** Increase penetration in the existing market with existing products. This can be done in the following ways: -
   - Attracting new prescribers
   - Increasing quantum of prescriptions from current prescribers by
     - o Prescribing to a greater number of patients
     - o Prescribing in more indications
   - Attracting prescribers from competitors

2. **Product development:** Improve existing products or introduce new products for existing markets.

   Improvements in existing products can be done for improving efficacy, reducing side effects, enhancing taste, convenient dosage, etc.

   The aim is to gain higher sales from the current market.

3. **Market development:** Existing products are sold in new markets. This can be done in the following two ways-

    Expand into new geographies

    Get into new market segments

4. **Diversification:** New products are developed for new markets. e.g., a pharmaceutical company marketing formulations diversifying into setting up hospitals or a pharmacy chain.

SUMMARY

This chapter deals with the new product concept, new product development process in pharma, pharmaceutical product life cycle, and product strategies for growth. Key points discussed in this chapter are as follows:

1. The new product could be new to the world, new to the company, adding to the existing product line, improvement, and revision of an existing product, repositioning or cost reduction of an existing product.

2. A new product launch is one of the important growth strategies. The majority of Indian companies have grown over the years by launching several new products which are "me-too" in nature. Big multinational companies have grown by launching their original research molecules in India.

3. Over the years, the number of new drugs discovered has come down primarily because new drugs take an average of $2 billion and more than ten years to reach the market after discovery.

4. New drug development is a complex process and it comprises various stages like

5. target identification and validation, lead identification and optimization, product characterization, formulation and development, preclinical research, investigational new drug filing, clinical trials, new drug application, and approval

6. The majority of new product launches in India are of generic drugs.

7. New generic product development process comprises of idea generation (identification of product), screening of products, sourcing of raw material, research & design, pilot batch manufacturing & conducting stability studies, drug master file (DMF), marketing strategy development, test marketing/clinical trial/user trial and commercialization

8. Pharma companies need to think of innovative strategies to succeed with the new product launches. Replicating successful launch strategies from the past is no longer a safe option since radical changes in physicians' preferences and behaviors have taken place due to the Covid-19 pandemic. Companies need to work on enabling the sales force to engage with doctors in innovative ways, analytics-enabled engagement with healthcare professionals, adopting innovative patient channels and services, providing personalized content digitally, adopting omnichannel distribution strategy, and exhibiting operational excellence.

9. As newer drug molecules/therapies are launched in the market they replace the older molecules/therapies. Thus, a pharmaceutical product passes through four stages of the life cycle - introduction, growth, maturity, and decline.

10. Product strategies for growth are market penetration, product development, market development, and diversification.

REFERENCES

1. New Products Management for the 1980s (New York: Booz, Allen & Hamilton, 1982).

2. Sahu Ram Prasad, Price drives Indian pharmaceutical market growth to a five-year high in FY20, Business Standard, Mumbai, May 11, 2020.

3. AWACS PharmaTrac MAT OCT 2021 data.

4. Ecem, Meşepınar, Özalp, Vayvay & Zeynep Tuğçe, Kalender. (2021). Management Of Idea-To-Product Process For Successful New Product Development In Pharmaceutical Industry. doi:10.15405/epsbs.2021.02

5. Nerkar, A., & Roberts, P. W. (2004). Technological and product-market experience and the success of new product introductions in the pharmaceutical industry. Strategic Management Journal, 25(8-9), 779-799. https://doi.org/10.1002/smj.417

6. Yousefi, N., Mehralian, G., Rasekh, H. R., & Yousefi, M. (2017). New Product Development in the Pharmaceutical Industry: Evidence from a generic market. Iranian Journal of pharmaceutical research: IJPR, 16(2), 834.

7. Anna Heid and Ivan Ostojic, "Recalculating the future of drug development with quantum computing," October 2020, McKinsey.com

8. Harputlugil Eylul, Hayton, Scott, Merrill Joey, and Salazar Pablo, First-time launchers in the pharmaceutical industry, McKinsey & Company, Feb 2021.

9. Deore, AB, Dhumane JR, Wagh HV, Sonawane RB, The Stages of Drug Discovery and Development Process. Asian Journal of Pharmaceutical Research and Development. 2019; 7(6):62-67, DOI: http://dx.doi.org/10.22270/ajprd.v7i6.616.

10. Lee Yeolan, Fong Eric, Barney Jay Bryan, and Hawk Ashton, Why Do Experts Solve Complex Problems Using Open Innovation? Evidence from the U.S. Pharmaceutical Industry, California Management Review 2019, Vol. 62(1) 144–166.

11. Thakur Vikram Pratap Singh, Ramacha Supriya, Pharmaceutical Business Strategy: A Generics Perspective, Journal of Intellectual Property Rights, Vol 17, Sept 2012, pp 484 - 496.

12. Rizwan Raheem Ahmed, Irfan Sattar, and Vishnu Parmar, Product Strategies in Pharmaceutical Marketing: A Perspective of Pakistani Pharmaceutical Industry, Middle-East Journal of Scientific Research 21 (4): 681-688, 2014.

13. Tamimi Nihad A.M., Ellis Peter, Drug Development: From Concept to Marketing! Nephron Clin Pract 2009;113:c125–c131.

14. Kotler P., Keller K., Koshy A., Jha M., Marketing Management: A South Asian Perspective,14th edition, Dorling Kindersley (India) Pvt. Ltd. 2013.

15. Ansoff H.I., Strategies for Diversification, Harvard Business Review, Sept-Oct 1957, 114.

CHAPTER 10

PRICING STRATEGIES

LEARNING OUTCOMES

After reading this chapter, you should be able to:

- Understand the objectives of pricing
- Understand the regulations regarding drug pricing
- Understand pricing methodologies
- Get an insight into the pricing strategies

PRICING

Price is the exchange value of a product or service. It is the only element of the marketing mix that generates revenue. Thus, it is important to set the right price for one's product or service.

Price serves several objectives; some important ones are stated below.

1. To maximize profits
2. To maximize unit sales
3. To gain a commanding market share
4. To discourage the entry of competitors
5. To create the perception of brand quality or exclusiveness
6. To increase trial purchases
7. To penetrate market
8. To recover the investment faster

PHARMACEUTICAL PRICING

Pricing of prescription drugs is relatively complex, involving issues related to intellectual property, doctors, and health insurance companies (Michael L. Capella et.al. 2009). In the pharmaceutical industry, competition depends on whether a brand-name drug has patent protection or whether both branded and generic versions of the drug are available. Even brand-name drugs under patent protection can face competition from other branded drugs that are considered therapeutic substitutes.

Within many drug classes, several possible substitute medications exist, and in such cases, the market structure is more appropriately depicted by the differentiated product oligopoly framework. Therefore, the optimal profit-maximizing price would equal marginal cost plus a positive margin, where the margin depends on the benefits and attributes (including prices) of the firm's own drug relative to other drugs in the class along with other available remedies. As a result, manufacturers price drugs according to the marginal value inherent in the pharmaceuticals such that prices of drugs representing important therapeutic gains may be greater than those of existing drugs used for the same purposes. The number of branded substitutes has a substantial negative impact on prices as well, reflecting competitive pressures (Berndt 2002, 2007).

Prices may be expected to have a greater influence on the choice of drugs prescribed by physicians. As physicians are increasingly competing for patients, they are trying to accommodate patients' price elasticity act in a price-sensitive manner even though they do not directly bear the cost of the drug (Gonul et.al. 2001).

Pharmaceutical pricing in India is highly regulated by the government. Healthcare expenditure has been a very important topic of discussion as every year about 50-60 million Indians fall into poverty due to the inability to finance health issues (The Hindu, Nov 19, 2020) since 70 percent of this is out-of-pocket expenditure (The New Indian Express,28 Feb 2021).

Of the total out-of-pocket healthcare expenses, a little over 50 percent is spent on purchasing medicines alone. The affordability of medicines is a crucial element in availing medical treatment by all sections of the people, particularly by the poor of the country. Price control has become an important tool in the government's hands to make drugs affordable. According to the data by AIOCD-AWACS, around 14 percent of drugs by value, and 25 percent by volume fall under price controls (V i s w a n a t h P i l l a, m o n e y c o n t r o l.c o m, N o v 0 5, 2 0 1 9).

REGULATIONS REGARDING DRUG PRICING

The government controls the prices of some bulk drugs and formulations. Drug Price Control Order (DPCO) is the regulation and National Pharmaceutical Pricing Authority (NPPA) is the regulatory authority.

NATIONAL PHARMACEUTICAL PRICING AUTHORITY (NPPA)

(Source: Annual Report 2020-21, Department of Pharmaceuticals, Government of India, Ministry of Chemicals and Fertilizers)

The National Pharmaceutical Pricing Authority (NPPA), an independent body of experts in the Ministry of Chemicals and Fertilizers, Department of Pharmaceuticals was constituted by the Government of India on 29.08.97. The functions of NPPA, inter-alia, include fixation and revision of prices of scheduled formulations under the Drugs (Prices Control) Order (DPCO), as well as monitoring and enforcement of prices. NPPA also provides inputs to the Government on pharmaceutical policy and issues related to affordability, availability, and accessibility of medicines. The Government notified DPCO, 2013 on 15th May 2013 in supersession of DPCO, 1995.

Salient features of DPCO, 2013 are as follows:

1. The National List of Essential Medicines (NLEM) notified by the Ministry of Health & Family Welfare is adopted as the primary basis for determining essentiality and is incorporated in the First Schedule of DPCO, 2013 which constitutes the list of scheduled medicines for the purpose of price control.

2. Ceiling prices of scheduled formulations are fixed based on 'market-based data'.

3. Price control is applied to specific formulations with reference to the medicine (active pharmaceutical ingredient), route of administration, dosage form/strength as specified in the First Schedule.

4. The National List of Essential Medicines 2015 (NLEM 2015) was notified by the Ministry of Health and Family Welfare in December 2015. It was, thereafter, notified as to the First Schedule of DPCO 2013, in March 2016, by the Department of Pharmaceuticals.

5. The functions of the National Pharmaceutical Pricing Authority (NPPA) are:

   - To implement and enforce the provisions of the DPCO, 1995 / 2013 in accordance with powers delegated to it.

- To undertake and/or sponsor relevant studies in respect of pricing of drugs/formulations.
- To monitor the availability of medicines, identify shortages, if any, and take remedial steps.
- To collect/maintain data on production, exports, and imports, market share of individual companies, the profitability of companies, etc. for bulk drugs and formulations.
- To deal with all legal matters arising out of the decisions of the Authority.
- To render advice to the Central Government on changes/ revisions in pharmaceutical policy.
- To render assistance to the Central Government in parliamentary matters relating to pharmaceutical pricing.

PRICE FIXATION

NPPA fixes the ceiling price of formulation listed in Schedule I of the DPCO, 2013. Under the market-based approach adopted in DPCO, 2013, the ceiling price of a scheduled formulation is determined by first working out the simple average of price to retailer (PTR) in respect of all branded-generic and generic versions of that particular formulation having a market share of one percent and above, and then adding a notional retailer margin of 16 percent to it. The maximum retail price (MRP) for that particular drug formulation must not exceed the notified ceiling price plus applicable taxes. NLEM 2015 contains 966 scheduled drug formulations (including formulations as per explanation 1 to Schedule – I of DPCO 2013) spread across 31 therapeutic groups. NPPA also fixes the ceiling prices of formulations listed under Explanation-I to Schedule – I of DPCO 2013. NPPA has fixed the ceiling prices of 881 formulations under DPCO, 2013 till 31st December 2020.

RETAIL PRICE

NPPA fixes the retail price of medicine based on the Form-I application received from the manufacturing/ marketing companies. The notified retail prices are applicable only to the applicant manufacturing/ marketing companies. The retail prices of the medicine are also fixed on the same method as applicable for the fixation of ceiling price.

Pricing determination

There are four basic approaches to pricing as follows: -

- Cost-based pricing

- Market demand-based pricing
- Competition-based pricing
- Market-based pricing

COST-BASED PRICING (ACCOUNTANTS' APPROACH)

This is an approach similar to the way an accountant would calculate the price for a product. It is based on the total cost of the product, including production and marketing costs, plus an allocation for overheads plus the target percentage to provide a profit margin. The total gives you a selling price. Problems with this approach are-

- Cost calculation is based on a predetermined level of demand and production. As these fluctuate, so does the product cost.
- It ignores market factors such as demand and competitors' actions.
- Overhead cost allocation can lead to a wrong pricing decision.
- A major benefit of this approach is that it can help indicate minimum price levels.

MARKET DEMAND-BASED PRICING (ECONOMISTS' APPROACH)

The aim of this approach is to explore the effect that different prices may have on the demand in the market for a product. Here you try to calculate the break-even point (produced by varying volume forecasts) based on different selling prices. This approach brings into focus the impact of price on volume and tries to find the most profitable price/volume ratio. You must ask yourself how many units of a given product you could sell at different price levels as explained in the example below: -

Price level	A Unit price (Rs.)	B No. of units sold	C Total revenue	D Fixed costs (Rs.)	E Variable costs	F Total costs	G Profit	H Break-even(units)
Calculations			A x B		B x Rs.4*	D+E	C-F	D/(A-4)
X	10	100	1000	100	400	500	500	17
Y	20	60	1200	100	240	340	860	6
Z	30	40	1200	100	160	260	940	4
*Assuming unit variable cost as Rs.4								

Fig. 10.1 Example of demand-based pricing

Let's see the meaning of each column: -

A. Price level – for example, we want to see the impact on profit and volume of setting the price at Rs.10, Rs.20, and Rs.30 per unit.

B. Number of units – these are estimated unit sales at each of the price levels set in A. If the price is Rs.10, one expects to sell 100 units. At Rs.30 per unit, only 40 units may be sold.

C. Total revenue is the total sales obtained by multiplying the number of units sold (B) by the selling price (A).

D. Fixed costs are sometimes referred to as 'overheads'. They include all costs which over a definite period of time (usually one year) do not vary with changes in the volume of sales. Examples are management salaries, factory and administration rent, heating, telephone, office expenses, fixed costs of sales-force salaries, cars, and so on.

E. Variable costs are those which vary according to the volume of product produced such as raw material, packaging, royalties, and commissions on sales. Total variable cost per unit in column E is calculated by taking the variable cost per unit and multiplying it by the number of units sold (B). The product Profit and Loss statements will give you information on product cost.

F. Total costs are simply the sum of fixed costs (D) plus variable costs (E).

G. Profit is calculated by subtracting total costs (F) from total revenue (C).

    As evident from the example above, if the assumptions of the price/volume relationship prove true, a maximum profit of Rs.940/- would be obtained at a price of Rs.30/- per unit.

H. Break-even consists of the minimum number of units that must be sold at the predetermined price and cost structure in order to cover all costs. At this break-even volume, given here in units, the profit is zero. Note that the lower the price is, the higher is the volume needed to break even concept can be conveyed graphically using the example.

The advantage of the market demand-based approach is that it brings together price calculations with market demand realities; that is, if demand for a product tends to be a function of its price, then this should be a determining factor in your decision.

The disadvantage is the difficulty of estimating the effect that price variations may have on product demand: one has to estimate how much one can sell in units for a given price level. Given this problem, an easy way to establish price elasticity is to examine the historical performance of similar products at a number of different price levels to study the effect on sales of price.

COMPETITION-BASED PRICING

The objective of this pricing approach is that it considers the prices set by competitors in the marketplace.   Competitive pricing can be approached in a number of ways:

- Prices can be set above your competitors' products.
- Prices can be set below your competitors' products.
- Prices can be set at the same level as your competitors.

You must also try to estimate your competitors' costs when considering their prices and developing your own pricing strategy. In this way, you will be in a better position to decide how much price latitude your competitors enjoy and whether they can use price against you aggressively.

Fig. 10.2 below offers a succinct format for gathering data for inclusion in your product plan. The basis of price comparison should be clearly indicated. Price comparisons between different products could be on a unit basis, on a daily-cost-of therapy basis (for pharmaceuticals), or on a price-per-procedure basis.

You should note the name of each competitive product being analyzed and the units' share of that product as well as yours. This will help you in determining the relationship that exists between price levels and share penetration. Here again, prices should be analyzed at all significant levels; that is hospital, wholesaler, public, pharmacy, and so on.

You should note the prices in the row indicated, and then calculate an index based on the company price for current products (obviously, for a new company product where the price is not yet set, you can use the market leader as a point of reference or the 'advised' price).

		COMPETITORS					
PRODUCTS:							
Units							
Mkt. Share							
Stockist	Price						
	Index						
Hospital	Price						
	Index						
Retailer	Price						
	Index						
Other	Price						
	Index						
Expected price %							
change for next year							

Fig. 10.2 Format for price comparison

MARKET-BASED PRICING

In this approach prices are based upon the 'value satisfaction' the product delivers to the buyer of the 'perceived' value. This 'perceived' value can be a result of:

- Value for money is influenced by the reputation of the firm, service level, the performance of the product (efficacy, ease of use, etc.).

- Image affected by status (endorsement by opinion leaders, exclusivity, or promotion).

- Reflection of different and distinctive market segments putting different 'value' on product performance.

- Price barriers are set in different segments such as government hospitals and private clinics.

The key to market pricing is to make an accurate assessment of a market's perception of the value of your product. Market research will be needed to avoid two dangers:

- Over-pricing because of your inflated view of the value of your product. Almost inevitably the 'perceived price' will be a qualitative judgment made by the buyer relative to his/her experience of the competition.

- Underestimating the real value and charging less than what you could.

PRICING STRATEGIES

1. **Price Leadership:** Decision-makers often tend to think of price ranges for products rather than the absolute price level.   If you can favorably distinguish your company product from a competitor, then you have positioned this product at the upper end of the segment. This 'valued' or favorable image must be reflected in an equivalent pricing policy. (Remember, to maintain this price you must keep reinforcing this image.)

   Given a price range, a decision has to be made as to where in that range to locate your product's price. This is a strategic decision made on the basis of corporate and company objectives, which may include some or all of the following:

   - Achieve target return on investment or sales.

   - Stabilize prices

   - Maintain or improve market share

   - Meet or prevent competition

   - Maximize profits

   There are a few important pricing strategies (skimming, penetration, marginal cost) which we will now review.

2. **Skimming pricing policy:** This strategy sets a price at the top of the acceptable price range. Skimming is often used:

   - On a new product in an early stage of the life-cycle to recoup high R & D investment.

   - To prevent pricing mistakes by being too low (it is easier to reduce than to increase prices if the wrong price level is chosen).

   Advantages

   - Used on a new product in the early stage of the life-cycle to recoup high R & D investment.

   - To segment the market.

   - To prevent pricing mistakes by being too low – it is easier to reduce than to increase prices if the wrong price level is chosen.

   - To limit off-take if plan capacity or stocks are not adequate.

Disadvantages

- Attracts competition
- Low volume may not suit plant output.
- Consumer awareness and acceptance will be slower in the introductory product life stage for a new product.
- More vulnerable to economic depression.

3. **Penetration pricing policy:** This strategy is the opposite of skimming. A low price is set, often below the existing range with the objective to gain maximum market penetration as quickly as possible; that is, low price, and high volume.

   Advantages of this strategy are –
   - Product economy of large-scale production.
   - Pre-empts competition.
   - Wins wide product allegiance for future; for example, after patent life.

   Disadvantages of this strategy are-
   - Profit return is lower and the pay-back period longer for a new product.
   - It will be disastrous if the product has a very short life cycle.
   - It can be difficult to overcome the psychological disadvantages of having to increase the price if the initial price was set too low.

4. **Marginal cost pricing:** In highly competitive situations one may have the opportunity of gaining business if a sufficiently low price is offered. This is especially true in hospital tendering. The question arises, however: what is the lowest price to use at which it makes sense to take the business? One approach is to use marginal costing which is defined as 'the cost of producing one more unit'.

   The cost of producing one more unit means that the fixed costs are already being covered by the existing sales volume, and then the costs of producing the extra unit are the variable costs. If a small profit is made per unit, then at least this is an additional contribution that would not have been there had we not obtained the extra business. It can be argued that, even at no profit, marginal business is worth having as it may use resources that would otherwise stand idle. Generating this type of business, however, will ultimately eat into profits and depress the percentage of return on sales. The major use of marginal costing, therefore, is to answer the question 'Should I accept this order?' rather than as a pricing tool. This type of strategy is more often used with price elastic, high-volume products, where it is important to keep the sales volume up.

SUMMARY

This chapter deals with pricing, regulations regarding drug pricing, pricing methodologies, and pricing strategies. Key points are as follows:

1. Price is the exchange value of a product or service. It is the only element of the marketing mix that generates revenue. Thus, it is important to set the right price for one's product or service.

2. Price serves several objectives like to maximize profits, to maximize unit sales, to gain commanding market share, to discourage entry of competitors, to create the perception of brand quality or exclusiveness, to increase trial purchases, to penetrate the market, and to recover the investment faster.

3. Pricing of prescription drugs is relatively complex, involving issues related to intellectual property, doctors, and health insurance companies. In the pharmaceutical industry, competition depends on whether a brand-name drug has patent protection or whether both branded and generic versions of the drug are available. Even brand-name drugs under patent protection can face competition from other branded drugs that are considered therapeutic substitutes.

4. Prices may be expected to have a greater influence on the choice of drugs prescribed by physicians. As physicians are increasingly competing for patients they are trying to accommodate patients' price elasticity act in a price-sensitive manner even though they do not directly bear the cost of the drug.

5. Pharmaceutical pricing in India is highly regulated by the government. Healthcare expenditure has been a very important topic of discussion as every year about 50-60 million Indians fall into poverty due to the inability to finance health issues since 70 percent of this is out-of-pocket expenditure. Of the total out-of-pocket healthcare expenses, a little over 50 percent is spent on purchasing medicines alone.

6. The government controls the prices of the number of bulk drugs and formulations. Drug Price Control Order (DPCO) is the regulation and National Pharmaceutical Pricing Authority (NPPA) is the regulatory authority.

7. The National List of Essential Medicines (NLEM) notified by the Ministry of Health & Family Welfare is adopted as the primary basis for determining essentiality and is incorporated in the First Schedule of DPCO, 2013 which constitutes the list of scheduled medicines for the purpose of price control.

8. Ceiling prices of scheduled formulations are fixed based on 'market-based data'.

9. Price control is applied to specific formulations with reference to the medicine (active pharmaceutical ingredient), route of administration, dosage form/strength as specified in the First Schedule.

10. There are four basic approaches to pricing - Cost-based pricing, Market demand-based pricing, Competition-based pricing, and Market-based pricing

11. Key pricing strategies include price leadership, skimming pricing policy, penetration pricing policy, marginal cost pricing

REFERENCES

1. The Hindu, Inability to finance health costs pushing 50-60 million Indians into poverty Nov 19, 2020.

2. Kumar Vikram, Nearly 70 percent expenditure on health to come out of patients' pockets: Finance Commission report, The New Indian Express, 28 Feb 2021

3. Viswanath Pilla, Explainer: How drug prices are regulated in India, moneycontrol.com, Nov 05, 2019, https://www.moneycontrol.com/news/business/explainer-how-drug-prices-are-regulated-in-india-4606751.html accessed on Nov 2, 2021

4. Annual Report 2020-21, Department of Pharmaceuticals, Government of India, Ministry of Chemicals and Fertilizers.

5. Capella Michael L., Taylor Charles R., Campbell Randall C., and Longwell Lance S., Do Pharmaceutical Marketing Activities Raise Prices? Evidence from Five Major Therapeutic Classes, Journal of Public Policy & Marketing Vol. 28 (2) Fall 2009, 146–161.

6. Berndt, Ernst R. (2002), "Pharmaceuticals in the U.S. Health Care: Determinants of Quality and Price," Journal of Economic Perspectives, 16 (4), 45–66.

7. Gonul, Fusun, Franklin Carter, Elina Petrova, and Kannan Srinivasan (2001), "Promotion of Prescription Drugs and Its Impact on Physicians' Choice Behavior," Journal of Marketing, 65 (July), 79–90.

CHAPTER 11

PHARMACEUTICAL DISTRIBUTION

LEARNING OUTCOMES

After reading this chapter, you should be able to:

- Understand the pharmaceutical value chain

- Understand the distribution channel structure

- Learn how to assess distribution channels

- Understand physical distribution and challenges

- Understand the hybrid distribution model

PHARMACEUTICAL VALUE CHAIN

Bulk drugs or active pharmaceutical ingredients (APIs) are manufactured by combining two or more chemicals or intermediaries (raw material). These are then used as raw materials for manufacturing finished products, which are ready-to-use dosage forms like capsules, tablets, syrups injections, etc. Through the distribution network, the finished products are then made available to the end consumers/patients. Marketing and Sales are done to capture the value. The pharmaceutical value chain is depicted in Fig.11.1.

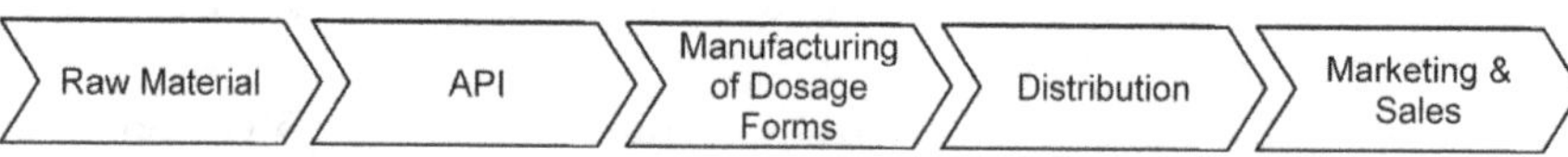

Fig. 11.1 Pharmaceutical Value Chain

Distribution: Distribution encompasses how manufacturers get their products and services into the hands of the consumer.   The objective of distribution is to get your products to the right place, in the right condition, and at the right time by the most cost-effective methods. Distribution is consequently one of the most vital as well as expensive tools of marketing. Generally, the ultimate consumer is in no position to go to your company directly, so a system has to be developed of getting the product to the consumer.   You need to understand two interrelated aspects of the distribution process:

Channel of distribution: These are intermediaries that perform a variety of functions between the manufacturer and the user/consumer of the product; for example, wholesalers, and retail pharmacies.

Physical distribution: The set of tasks involved in physically moving products from the original manufacturer to the points where the product is used or consumed to meet the needs of the consumers; for example, warehousing, order processing, outbound transportation, customer delivery, repeat servicing of hospitals, pharmacies, and so on.

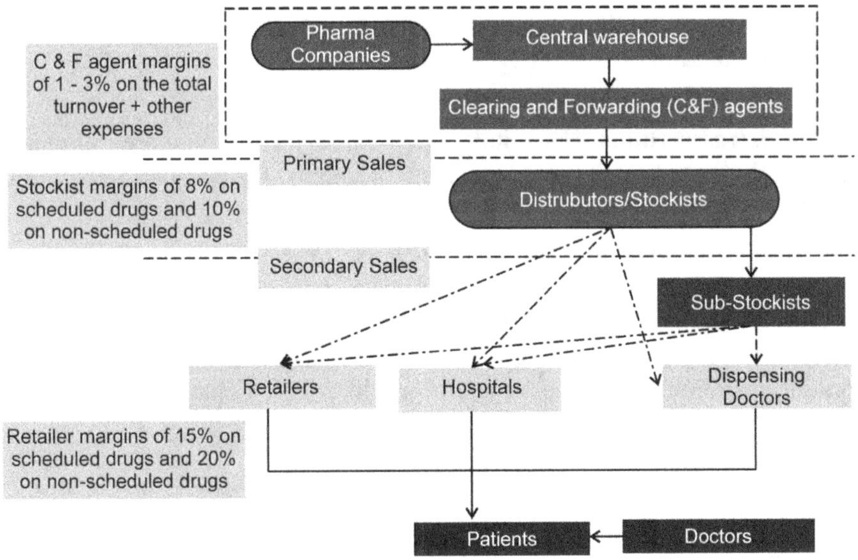

Fig. 11.2 Pharmaceutical Distribution Channel

Distribution is a vital link in the sale of your products to the consumer. When you promote your product and convince the doctor to prescribe it or the hospital to use it, that product has to be immediately available where it is needed. Without effective and efficient distribution channels you have lost control over a vital part of your marketing mix by making it difficult for a customer to buy your product.

Channels of Distribution: The pharmaceutical distribution in India is highly fragmented. It comprises Clearing & Forwarding Agents (CFAs), Stockists/Distributors, Retailers, and Hospitals as depicted in Fig 11.2.

CFAs: These organizations are primarily responsible for maintaining, the storage (stock) of the company's products and forwarding drugs to the stockist on request. Most companies keep 1–3 CFAs in each Indian state. On average, a company may work with a total of 25–35 CFAs. The CFAs are paid by the company yearly, once or twice, based on the fixed percentage of the total turnover of products.

Stockist: Is the distributor, who can simultaneously handle more than one company (usually 5–15 depending on the city area), and may go up to even 30–50 different manufacturers. They pay for the products directly in the name of the pharmaceutical company after 30 to 45 days. Stockists get margins of 8% on scheduled drugs and 10% on non-scheduled drugs.

The Retail Pharmacy: Obtains products from the stockist or sub-stockist through whom it finally reaches the consumers (patients). There are about 8,50,000 retail pharmacy outlets in India. Retailers get margins of 16% on scheduled drugs and 20% on non-scheduled drugs.

Hospitals/Institutions: Hospitals and other institutions are supplied through dedicated institutional stockists or agents. Certain big institutions could be supplied directly from the company CFA.

Assessing Distribution Channels: Because of their potential impact on the success of your product's performance, you need to assess each of your distribution channels so that you can identify where improvements can be done or, if you are able, find another shorter more efficient channel to your customers. To assess the current distribution channels, you could review the following:

- Total sales of your product to all channels.
- Total sales to each channel.
- Invoice sizes – any very large invoices to one customer that are the exceptional indicating product being shipped in one drop to say a major hospital or pharmacy.
- Levels of monthly purchases
- Out-of-stock frequency.
- Delivery times.

- Invoice systems/times.

- Credit terms

- Return-of-goods procedures/policy/frequency.

- How large is an inventory normally carried?

- Bonus schemes, incentives taken advantage of in the past, and the knock-on effect in sales, if any.

Finally do not forget that the sales force provides another source of information about channels of distribution. But if you are going to use this source, structure the information you ask them to collect from wholesalers, hospitals, and retail pharmacies so that it is consistent and not just a collection of opinions passed on. Remember, sales personnel are deliberately biased. They are persuasive advocates of your products.

Although it is not easy to get absolute numbers regarding levels of sales by distribution channels, it is possible to compare alternative channels using the following equation:

$$\text{Rate of return} = \frac{\text{Estimated sales volume} - \text{Cost of channel}}{\text{Cost of channel}}$$

The purpose of this calculation is to give you an idea of which distribution channel produces the best rate of return and therefore should be emphasized in promotion, service, and support.

Physical Distribution: Physical distribution, starting with the products at the manufacturing point or plant, is concerned with the most cost-effective methods of getting them to the consumer/user. Whilst the product manager is not normally responsible for the physical distribution of his particular products, there are several aspects that he needs to be involved in to ensure that his product plans are carried through as planned. These are:

- Speed and efficiency in receiving, processing, and delivering orders: Average minimum/maximum times; level of complaints, from whom, how dealt with, and so on.

- Warehousing capacity at the company and wholesalers: Was it adequate last year?

- Will it be adequate to meet marketing plans?

- Inventory levels: Do your inventory levels fall below demand, how often, for how long, and what remedial actions are taken then? Have wholesalers/retail pharmacy levels fallen in the last year(s)?

What effect has there been on your ability to supply and at what costs?

- The company's ability to meet emergency deliveries of products to customers: Who requested them? How often? Why? How quickly does your company deliver? At what cost over budgets?

- Quality control: Complaints, how many, about what, what actions were taken?

- Company policy/willingness to take back defective or damaged products and resupply quickly: What complaints have been received in the last financial year? How do you compare with competitors? Has any competitor's policy in this area led to your product not being stocked?

Challenges: The distribution in the pharmaceutical industry is highly fragmented with various stakeholders such as CFAs, stockists, hospital stockists, generic pharma distributors, etc. which makes it very complicated. There are numerous brands and multiple SKUs (stock-keeping units) of each brand which further complicates the entire system. Companies lose visibility and control of the movement of the product beyond the CFAs. To overcome external infrastructure constraints, the pressure for timely and effective supply chain execution becomes paramount. Also, the non-availability causes serious threats of counterfeits, loss of sale, challenges to trace products, and unpredictable demand scenarios. Moreover, reverse logistics takes a toll on costs that are adjusted in regular sales. Cold chain maintenance for specialty products like biologics adds another challenge given the diverse environmental condition in India. A temperature-controlled supply chain adds a burden to the cost of distribution and profitability (**Amaninder Dhillon,** Indiamedtoday, March 25, 2021). New models, of information technology solutions such as RFID, enabled drug anti-counterfeiting, vendor-managed inventories, hospital inventory management, patient-centric digital homecare solution, end-to-end supply chain visibility of demand forecasting and analytical insights, blockchain with predictive analysis, digital supply chain solutions will be the next disruptive models.

Hybrid Distribution Model: The Covid-19 pandemic has given a boost to the online purchase of medicines. The e-pharmacy is expected to grow strongly, and the overall number of households served is set to cross the 70 million mark by 2025(https://ehealth.eletsonline.com/2021/04/e-pharmacy-the-next-big-scope-in-pharmabiz/). The pandemic demonstrated how the physical and online pharmacies need to work in tandem and both play a crucial role in improving access to quality and affordable medicines across the country. Today multiple online companies with their hub-spoke model support more than 10000+ pin codes across the country with the help of supply chain partners and last-mile delivery companies. Increasing digitization has in turn led to increasing e-commerce adoption. Furthermore, growing awareness among the population concerning counterfeit drugs is driving people towards organized and digital channels offering medicines, thereby supporting the market growth over the next few years.

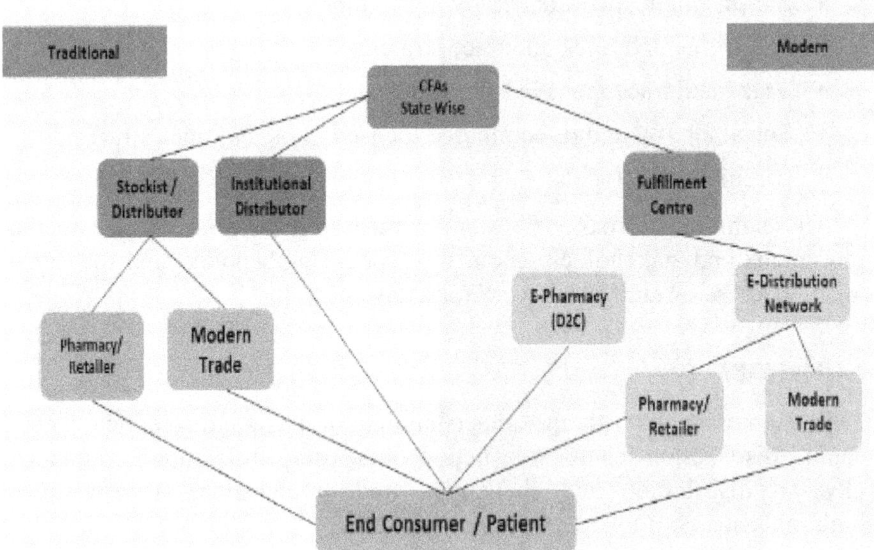

Fig. 11.3 Hub and Spoke Model for e-commerce, e - pharmacy, e-distribution

Blockchain authentication against counterfeits: The pharmaceutical industry is growing at a rapid pace, with large investments and demand for products not only in India but across the globe. As the brands in pharmaceuticals grow by value and volume, it is evident that the counterfeiting industry makes its way to cull part of the profit and extract money from the legitimate supply chain. To counter this, technology companies have come up with multiple solutions from a simple QR bases authentication system to complex blockchain methods. The blockchain method is interesting as pharma manufacturers can now also authenticate APIs, Excipients, and other relevant ingredients to their source and date of manufacturing. This involves supply chain embedded parameters to track and trace product origins.

AI-based supply chain initiatives: With AI-based prediction models now embedded in the current generation ERP systems, pharmaceutical companies can now do the following:

Visibility of products across the supply chain

- Counterfeit avoidance strategies
- Cost efficiency in supply chain operations
- Demand forecasting for optimizing production
- Supply chain software deployment support
- Packaging and track and trace strategies
- Track and trace managed services
- Sustainability and re-commerce focused track and trace strategies
- Circularity protocols

These initiatives make it easier for marketers to communicate to the physicians and stakeholders about the product authenticity and measures taken to keep it watertight (https://tongadive.com/).

SUMMARY

This chapter deals with the understanding of the pharmaceutical value chain, distribution channel structure, assessing distribution channels, physical distribution and challenges, and omnichannel strategy. Key highlights are as follows: -

1. Distribution encompasses how manufacturers get their products and services into the hands of the consumer. The objective of distribution is to get your products to the right place, in the right condition, and at the right time by the most cost-effective methods.

2. The pharmaceutical distribution in India is highly fragmented. It comprises Clearing & Forwarding Agents (CFAs), Stockists/Distributors, Retailers, and Hospitals.

3. To assess the distribution channels, you need to review total sales of your product to all channels, total sales to each channel, invoice, levels of monthly purchases, out of stock frequency, delivery times, invoice systems/times, credit terms, return-of-goods procedures/policy/frequency, how large an inventory is normally carried, bonus schemes, incentives.

4. Physical distribution, starting with the products at the manufacturing point or plant, is concerned with the most cost-

effective methods of getting them to the consumer/user. It involves transportation, inventory management, warehousing, quality control, stocks-return policy, etc.

5. With the growth in online pharmacy hybrid distribution models of physical as well as digital will continue to grow.

6. Blockchain, AI-based technology will be widely applied in the future to prevent counterfeiting and to bring cost efficiency in distribution.

REFERENCES

1. Dhillon Amaninder, Reshaping pharma supply chain, Indiamedtoday, March 25, 2021, https://indiamedtoday.com/reshaping-pharma-supply-chain/ accessed on April 2, 2022.

2. https://ehealth.eletsonline.com/2021/04/e-pharmacy-the-next-big-scope-in-pharmabiz/

3. https://tongadive.com accessed on April 2, 2022

IMPORTANCE OF RETAIL PHARMACY & EMERGENCE OF E-PHARMACY

LEARNING OUTCOMES

After reading this chapter, you should be able to:

- Know about the pharmaceutical retail universe

- Understand the growing importance of retail

- Understand the growing influence of e-pharmacy

- Get an insight into e-pharmacy advantages and challenges

PHARMACEUTICAL RETAIL UNIVERSE

The pharmaceutical retail universe in India comprises unorganized players (independent pharmacy outlets), organized players (pharmacy outlet chains like Apollo pharmacy), and online retail. With around 8,50,000 pharmacy stores spread across the country, retail pharmacies are still the dominant medical distribution channels in India. Offline drug stores account for over 85 percent of the overall pharmaceutical sales and include the sale of over-the-counter drugs, prescription drugs as well as a variety of FMCG products (https://www.statista.com/statistics/1027563/india-retail-pharma-market-size/).

THE GROWING IMPORTANCE OF RETAIL PHARMACY

Increased affordability, improved accessibility, and epidemiological transition are expected to fuel the pharmacy retail market in India. The growing incidence of chronic diseases like hypertension, diabetes, cardiovascular diseases, and cancer, which require prolonged medication, is expected to further drive the growth of the pharmacy retail market in India.

The rise in per capita income and penetration of health insurance coverage has helped propel consumer spending on pharmaceuticals. With the rise in public healthcare expenditure, pharmaceuticals have become more accessible, thereby fueling the pharmacy retail market, especially in rural areas and tier-II cities. To make medicines affordable to all, the government launched the Jan Aushadhi scheme in 2008 which was renamed "Pradhan Mantri Bhartiya Janaushadhi Pariyojana" (PMBJP) in November 2016. There are 8012 Jan Aushadhi kendras across the country as of March 2022 (http://janaushadhi.gov.in/).

THE GROWING INFLUENCE OF e-PHARMACY

The e-Pharmacies are online platforms where consumers can purchase medicines without having to visit brick-and-mortar pharmacies. This makes the process more convenient for consumers and has resulted in a rising demand for the model across the world. According to the HTF Market Intelligence report titled 'E-Pharmacy Market in India 2022', the e-pharmacy market was valued at INR 50.71 Bn in 2020 and is estimated to reach INR 458.14 Bn by 2026, expanding at a compound annual growth rate (CAGR) of ~44.99% during the 2021 – 2026 period.

The Covid-19 pandemic has given a boost to the online purchase of medicines. The e-pharmacy is expected to grow strongly, and the overall number of households served is set to cross the 70 million mark by 2025(https://ehealth.eletsonline.com/2021/04/e-pharmacy-the-next-big-scope-in-pharmabiz/). The pandemic demonstrated how the physical and online pharmacies need to work in tandem and both play a crucial role in improving access to quality and affordable medicines across the country. All over the world, wherever e-pharmacies operate, they co-exist with retail pharmacies. While e-pharmacies are viable due to their focus on chronic therapies, retail pharmacies are necessary for the immediate requirement of acute medicines such as pain killers and antibiotics.

Factors driving the demand for e-pharmacy in India are as follows: -

- Increasing penetration of the internet and smartphones in both urban and rural India.

- Increasing digitization has in turn led to increasing e-commerce adoption. Also, changing consumer preferences towards online shopping has led to widespread penetration of various e-commerce channels and pharmacies selling medicines online.

- A growing number of chronic elderly patients living in nuclear families, and patients who are not in a condition visit a pharmacy outlet.

- Technology advancements are coming up in the form of applications that help in bringing price transparency, creating awareness, finding an appropriate healthcare service provider, medicine reminders, and pregnancy alerts to the consumers.

- Issues related to retail pharmacy like tracking authenticity, traceability of medicine, abuse prevention, addressing consumption of drugs without prescription

Today multiple online companies with their hub-spoke model support more than 10,000+ pin codes across the country with the help of supply chain partners and last-mile delivery companies.

The ever-moving landscape of mergers and acquisitions in the e-pharmacy and e-commerce space has made it an exciting industry to be associated with. The major e-pharmacy players in India today are (as of March 2022):

- PharmEasy
- 1Mg
- Apollo (Hybrid)
- Netmeds
- Amazon (pharmacy)
- Flipkart (pharmacy)
- Practo
- Lyberate
- Docprime
- mChemist
- Medplusmart (hybrid)
- Wellness Forever (hybrid)

Furthermore, growing awareness among the population concerning counterfeit drugs is driving people towards organized and digital channels offering medicines, thereby supporting the market growth over the next few years. In terms of the product offering, the following are products offered:

1. Prescription Medication
2. OTC
3. Cosmetics
4. Nutraceuticals
5. Medical Foods
6. Sports Nutrition

THE E-PHARMACY MODEL

An e-pharmacy model is required to have two operating components for dispensing prescription medicines (FICCI and Frost & Sullivan report on E-PHARMACY IN INDIA: LAST MILE ACCESS TO MEDICINES):

1. **Technology:**
   - Web-based and/or mobile-based application for consumers to upload the scanned copy of their prescriptions and place requests for medicines
   - Every order that is received to be verified and checked by a team of registered pharmacists
   - The registered pharmacists forward the validated prescriptions to the pharmacy store from where the medicines are dispensed
   - The web or mobile-based platform to be governed under the IT Act 2000 and only act as a platform to facilitate a connection between consumer and pharmacy store

2. **Pharmacy Retail Store:**
   - The licensed pharmacists of the store check for the validity of the prescriptions, failing which the medicines would not be dispensed
   - The medicines should be dispensed from a licensed premise in a sealed tamper-proof pack to the patient or patient's relative
   - There should be a proper invoice with the batch number of the medicines dispensed, expiry date, name and address of the pharmacy with the signature of the registered pharmacist/s

- The pharmacy store is to be operated under the oversight of the Drugs and Cosmetics Act & Rules and needs to comply with all the requirements of the act, as it does for its normal business

BENEFITS OF E-PHARMACY

A. For Patients / Consumers

- Consumers can conveniently order medicines from their mobile phones or computers. This will significantly help patients who are old and sick and not in a condition to go out to a retail pharmacy.

- With the use of technology and access to an inventory of multiple stores at a time, e-pharmacies can aggregate supplies, making otherwise hard-to-find medicines available to consumers across the country. Retail pharmacies can only keep a limited inventory, resulting in the consumer having to visit multiple stores to procure the medicines.

- Enable access to rural areas where there is limited presence of retail pharmacy.

- The consumer can choose from a wide range of affordable generic equivalents for a particular branded drug, which is not possible in the current offline model.

- Cost advantage to end consumers as e-pharmacies can broaden their customer base while reducing working capital, overhead costs, and increasing margins

- Information to consumers on drug interactions, side effects, medicine reminders, and cheaper substitutes can be provided through e-pharmacies. This can help in improving patient compliance.

- All medicine purchases are digitally stored making it easy to track the supply chain, thereby decreasing the risk of counterfeit medicines, drug abuse, and self-medication.

B. For the Regulators

- All the transactions could be efficiently tracked with complete details of the medicines, batch number, dispensing pharmacy name and address, prescribing doctor, name and address of the patient, etc., thereby reducing the problem of drug abuse and self-medication.

- The e-pharmacies store and analyze large amounts of data on consumers across the nation, which would be useful for planning public health policies.

- The technology-enabled tracking systems of the e-pharmacy model assist in back-tracing the channel/manufacturer/supplier of the counterfeit medicines, thereby making the market a lot more transparent and authentic.

- Since the e-Pharmacy model has a stringent documentation process, the taxes paid on all transactions will largely benefit the Government.

C. For the Pharmacists

- An e-pharmacy model can enable existing pharmacies to start online operations and serve a broader set of customers or a network of pharmacies.

- An e-pharmacy model ensures consolidation of inventory thereby reducing the working capital requirements, removing wastage from the system, and increasing margins, thus making the e-pharmacy, a sustainable value-added service model.

- The e-pharmacy model enhances the services of the pharmacist to the consumers like answering questions about medications using e-mail or other real-time chat options.

D. For Pharmaceutical Companies: Companies can track sales of individual products so they can accordingly tailor their efforts & investments

- Able to track detailed demand of products at a pin-code level, which directly informs sales & marketing interventions targeted at both consumers & doctors levels.

- In the case of e-pharmacies patients tend to stick to the prescribed therapies ensuring fewer medicine drop-offs compared to traditional pharmacies.

- Since e-pharmacies have a much wider consumer reach it provides a great marketing channel with the ability to generate demand for new products

Challenges

- Medicine is one of the products which customers want on priority, especially in acute conditions. So, delivery is always going to be challenging. Currently, most e-commerce giants play on this point and delivery is up to 3 hours in most cases.

- Most of the pin codes have local medical stores so pin code coverage is extremely important to improve the market share.

- To fulfill the above two challenges, e-pharmacy players need to have small or big warehouses depending on the market so it increases the costs of logistics, rent, etc.

Future: Many e-pharmacies adopted an aggregator-based business model in which they relied on physical pharmacy retailers for their last-mile delivery to the customer's doorstep. Thus, e-pharmacies played a complementary role, which means online and offline stores will coexist. Chances are that online e-pharmacies may set up their stores due to which they will be able to deliver medicines faster and cheaper as compared to local medical stores.

All the e-pharmacy companies are not only selling medicines. They are selling multiple categories just like medical stores do and eventually they may introduce their private labels which will compete with the other companies.

SUMMARY

This chapter deals with the pharmaceutical retail universe, the growing importance of retail, the growing influence of e-pharmacy, e-pharmacy advantages, and challenges. Key highlights are as follows:-

1. The pharmaceutical retail universe in India comprises unorganized players (independent pharmacy outlets), organized players (pharmacy outlet chains like Apollo pharmacy), and online retail. With around 8,50,000 pharmacy stores spread across the country, retail pharmacies are still the dominant medical distribution channels in India.

2. Increased affordability, improved accessibility, and epidemiological transition are expected to fuel the pharmacy retail market in India. The growing incidence of chronic diseases like hypertension, diabetes, cardiovascular diseases, and cancer, which require prolonged medication, is expected to further drive the growth of the pharmacy retail market in India. Increased penetration of health insurance coverage has helped propel consumer spending on pharmaceuticals.

3. The e-Pharmacies are online platforms where consumers can purchase medicines without having to visit brick-and-mortar pharmacies. This makes the process more convenient for consumers and has resulted in a rising demand for the model across the world. According to the HTF Market Intelligence report titled 'E-Pharmacy

Market in India 2022', the e-pharmacy market was valued at INR 50.71 Bn in 2020 and is estimated to reach INR 458.14 Bn by 2026, expanding at a compound annual growth rate (CAGR) of ~44.99% during the 2021 – 2026 period.

4. The e-pharmacy model provides several benefits to patients/consumers like the convenience of buying, cost advantage, wide range, access in rural areas, information on medicines, authenticity, etc.

5. Benefits to regulators are tracking of data which can be used in framing policies, reducing drug abuse and counterfeit medicines.

6. Benefits to pharmacists are the opportunity to grow business, reduce cost and enhance services to patients/consumers.

7. Benefits to pharmaceutical companies are data tracking can help in decision-making about marketing efforts and investments. Also, it provides a marketing channel to generate demand for new products.

8. Challenges are in terms of pin code coverage, ensuring fast delivery, etc.

REFERENCES

1. https://www.statista.com/statistics/1027563/india-retail-pharma-market-size/

2. http://janaushadhi.gov.in/ accessed on April 2, 2022.

3. https://www.htfmarketreport.com/reports/3624964-e-pharmacy-market-in-india-2021 accessed on April 22, 2022.

4. https://ehealth.eletsonline.com/2021/04/e-pharmacy-the-next-big-scope-in-pharmabiz/

5. FICCI and Frost & Sullivan report on E-PHARMACY IN INDIA: LAST MILE ACCESS TO MEDICINES.

<h1>C H A P T E R 13</h1>

BUILDING PHARMA BRANDS

LEARNING OUTCOMES

After reading this chapter, you should be able to:

- Understand the significance of brand building

- Understand the concept of brand equity

- Understand brand-building process

- Learn how to track brand health

- Understand brand manager's roles and responsibilities

- Understand the brand planning process

BRAND BUILDING

Brands are the biggest assets that any company can have. Brand building refers to building brand value and a lasting and defensible competitive advantage for companies. A pharmaceutical company deals with several stakeholders like healthcare professionals, patients, retail pharmacies, distributors, suppliers of raw materials, government, and media. A brand that has value gains greater commitment from all of them. A strong brand is the source of strength for an organization. It helps the company to generate greater margins and better profits. For pharmaceutical companies, a patent is a fundamental tool, as it protects innovations and numerous investments in R&D to achieve high profits, but these rights conferred by a patent can have a maximum duration of twenty years. Instead, the value generated by a strong brand has an indefinite duration.

Consumer goods companies in India have been focusing on building brands. If we consider top FMCG brands in India, all of them have several brands with a turnover of over INR 1,000 crores. On the other hand, if we look at pharmaceutical companies, there is not a single brand with a turnover above INR 1000 crores. The Indian pharmaceutical industry is crowded with branded generics. Pharmaceutical companies have been focusing more on launching new products than on building brands. Pharmaceutical companies have grown over the years by launching new products and increasing doctor coverage by expanding the field force. However, with the new product pipeline drying up, increasing competition, and reaching a threshold of field force expansion, companies are now focusing on building brands.

BRAND EQUITY

There are several definitions of Brand Equity. Some of the key definitions are as follows: -

Brand equity is referred to as the value that consumers associate with a brand (Aaker, 1991). It's the consumers' perception of the overall product's excellence, carrying that brand name when compared to other brands. The value consumers associate with a brand is reflected in the dimensions of brand awareness, brand associations, perceived quality, brand loyalty, and other proprietary brand assets (Aaker, 1991).

High brand equity levels are known to lead to higher consumer preferences and purchase intentions (Cobb-Walgren et al., 1995).

The set of associations and behaviours on the part of the brand's consumers, channel members, and parent corporation permits the brand to earn greater volume or greater margins than it would without the brand name and that gives the brand a strong, sustainable, and differentiated advantage over competitors (Leuthesser, 1988).

BRAND BUILDING PROCESS

Stages in the brand building are as follows (adapted from Aaker, 1991; Keller, 1993 and PharmaVOICE, March 2018): -

A. Clearly define your brand purpose and the objective

When you want to define your brand, you'll need to ask yourself some questions:

- What problem does my brand solve?

- Who is my ideal customer?

- Who is my competitor?

- What are my customers' pain points?

- How does my brand make my customers feel?

- Why do my customers trust me?

- What is the story behind why my brand was created?

- If my brand was a person, what would their personality be like?

B. **Identify customer insights and define customer persona:** It is important to understand who your audience is and clearly define the customer persona. A customer persona is a detailed description of the ideal prescriber/buyer who would not be able to resist prescribing/buying your product. This persona helps you understand what type of person you are targeting with your marketing. You can't have an emotional impact if you don't know who it is you are trying to connect to.

C. **Brand identity and brand awareness:** It is the brand elements like name, logo, color, symbol, tagline, packaging, etc. that give your brand its identity. Brand awareness refers to how easily the target audience can recall your brand and recognize your brand. Brand awareness involves linking the brand to name, logo, symbol, etc. to certain associations in memory. Creating brand awareness involves giving the product identity by linking brand elements to a therapeutic category and associated disease condition.

D. **Brand positioning, values, and personality:** Please refer to chapter 7 on segmentation, targeting, and positioning.

Building the pharmaceutical brand strategy requires careful thought to be given to the brand position, values, distinguishing features, personality, and how the audience will view and receive the brand. When it comes to positioning the brand, marketing teams need to consider where the product sits in its therapeutic category, the various needs of the patients who will use it and the physicians who will prescribe it, which segment is most likely to benefit from it and respond to the product and determine how the brand will benefit those individuals, and assess customer response to the brand and, if necessary, refine the brand message.

The brand personality is an important element in positioning a product and differentiating it and can help to build loyalty and trust.

Giving brands human characteristics can help companies build relationships with customers and, ultimately, transform sales.

Brand values need to go beyond the traditional message of functional attributes — messages that are important to a physician but don't necessarily differentiate a product in a therapeutic class — and include the value they bring to support the needs of the patient. The message might be how quickly a product works to improve the patients' day-to-day life or convenience in how a drug is administered. Those brand values are key to building brand loyalty.

E. **Developing a promotion mix that would build trust and loyalty:** Please refer to chapter 14 on integrated marketing communications.

Trust is fundamental in building strong brands. Execute a promotional plan that would build trust amongst doctors, patients, caretakers, patient advocates, etc. Innovation in areas like experiential reality - augmented reality, mixed reality, and virtual reality is opening doors for brands to create more meaningful and memorable types of customer experiences. With strong medical storytelling, one can leverage experiential reality to help healthcare professionals get a deeper understanding of what you're trying to communicate.

A doctor's brand loyalty often depends on how the brand reinforces the doctor's sense of self and personal brand of medical practice, and not just the scientific facts.

F. **Commit to long term:** Initiate long-term programs that add value beyond the product and show commitment to the community and patients.

TRACKING BRAND HEALTH

Brand health tracking is more than just sales tracking. It involves the following analysis: -

1. Sales analysis - SKU-wise, region-wise, territory-wise, productivity, etc.

2. Market share analysis - within the same therapeutic category as well as other therapeutic categories prescribed by doctors for the similar purpose

3. Prescription analysis - specialty -wise prescription share, therapeutic category prescription share, indication -wise prescription share, etc.

4. Check whether segmentation, targeting, and positioning are right.

5. What are customer perceptions about the brand?

6. Is brand communication aligned to brand strategy?

7. Brand drivers - what has contributed to brand growth

8. Brand inhibitors - what has slowed down the brand growth

9. Does the brand manager have a thorough understanding of the brand, customer, and the market?

10. Brand portfolio analysis

11. Is investment in brand promotion and innovations in brand extensions enough?

12. Market analysis - key competitors and their market shares, revenues, etc. in comparison with the brand

13. Competitor analysis - key competitors and their promotional strategies versus brand strategies

ROLES AND RESPONSIBILITIES OF A BRAND MANAGER

The brand Manager is responsible for the following: -

- Brand Building

- Create impactful communication and campaign to build brands and therapies in the Division

- Improve Specialty Coverage and Therapy Focus

- Analyze market conditions and forecast sales and market share

- Implement Innovative and Creative Business Solutions to grow the market/prescription base

- Work effectively with creative agencies as partners, to develop differentiated and impactful brand campaigns

- Ensure quality and timely delivery/execution of campaign collaterals/inputs/activities to field force

- Continually incorporate customer feedback and macro view of the market through formal sources of feedback like research (Structured / MR driven etc.)

- Conduct research as per the brand need (analyze customer behavior, perception- attribute mapping, etc.) and design strategies accordingly. Undertake feasibility analysis/ market research for new opportunities.

- Develop and implement an Annual Operating Plan for the brand portfolio
- Undertake fieldwork to establish a strong connection with KOLs
- Actively drive engagement with sales team - conduct sales meetings to drive excitement
- Ensure quality and timely delivery/execution of campaign collaterals/inputs/activities to field force
- Effectively collaborate with cross-functional stakeholders for a buy-in of strategies to internal customers and ensure strong implementation of designed strategies
- Collaboration with the sales team for effective and result oriented execution

BRAND PLANNING PROCESS

Brand planning is to review your brand's performance, based upon metrics and insights gained, make any large strategic adjustments for the new year, and begin building a plan. This needs a strong understanding of the market and customer insights that will lead to a behavior change, and you still need a strong strategy to execute against those insights.

The brand planning process involves the following stages:

1. **Situation analysis:** This consists of the following-

   A. Market analysis
   - Market Progression (value and volume): size and growth
   - Correlation of trends with any key events: internal/external
   - Market forecasts
   - Key market drivers and trends
   - Total number of patients suffering from a particular disease

   B. Brand performance analysis
   - Sales achievement v/s budget
   - Market share achievement
   - Growth achievement v/s budget
   - SKU analysis
   - Regional analysis
   - Productivity per field force
   - Prescriber base

- Per capita prescription
- Brand share in Target Indication
- Brand Share in Target Specialty
- Brand Share of Voice

C. Competitor analysis

For key competitors provide-

- Strategic framework - strategy, key message, promotional spend, field force
- Effort benchmarking - doctor coverage, promotion mix
- Tactical framework - customer engagement strategies, scientific education, branding
- Relative competitive advantage - our brand advantage over competitors

D. Environment scanning

- Political
- Economic
- Regulatory
- Social
- Technological
- Reimbursement

2. **SWOT analysis:** Strengths - core competencies, unique resources, better than others in certain areas, cost advantage, what people in the market see as its strength

   Weaknesses - what can the organization improve, in what is it lagging behind the competition, in which areas fewer resources than competitors, what people in the market see as its weakness

   Opportunities - favorable market trends, growing customer demands, opportunities open to the organization

   Threats - competition, disruptive technology, obstacles to growth

3. **Brand Strategy:** This consists of –

   - Brand vision, values, and purpose
   - Brand objectives - revenues, market share, growth, ranking, productivity,   profits
   - Segmentation, targeting, and positioning

4. Creating single-minded proposition

- Identify customer insights
- Identify rational and emotional product benefits
- Identify users and brand adjectives
- Create a brand essence
- Write down the brand positioning statement

5. Marketing tactics

- Promotional plan - month-wise key initiatives
- Sales initiatives - month-wise key initiatives
- Total promotional cost on marketing and sales initiatives

6. Brand Profit & Loss statement

- Net sales
- Cost of goods
- Contribution
- Fixed cost
- Sales & admin cost
- Marketing cost
- Distribution cost
- Tax
- Profit

7. Monitoring & review:

- How are you going to monitor the execution of the plan?
- Reviews - monthly, quarterly, half-yearly

8. Contingency plan

- A contingency plan is a plan devised for an outcome other than expected
- Typically put in place for variations from key assumptions used in developing brand strategy and forecast

SUMMARY

This chapter deals with an understanding of brand equity, brand building process, tracking brand health, the role and responsibilities of a brand manager, and the brand planning process. Key highlights are as follows: -

1. Brands are the biggest assets that any company can have. Brand building refers to building brand value and a lasting and defensible competitive advantage for companies.

2. Pharmaceutical companies have been focusing more on launching new products than on building brands. Pharmaceutical companies have grown over the years by launching new products and increasing doctor coverage by expanding the field force. However, with the new product pipeline drying up, increasing competition, and reaching a threshold of field force expansion, companies are now focusing on building brands.

3. Brand equity is referred to as the value that consumers associate with a brand. It's the consumers' perception of the overall product's excellence, carrying that brand name when compared to other brands. The value consumers associate with a brand is reflected in the dimensions of brand awareness, brand associations, perceived quality, brand loyalty, and other proprietary brand assets. It is the set of associations and behaviours on the part of the brand's consumers, channel members, and parent corporation that permits the brand to earn greater volume or greater margins than it would without the brand name and that gives the brand a strong, sustainable, and differentiated advantage over competitors.

4. The brand-building process involves clearly defining your brand purpose and the objective, identifying customer insights and defining customer persona, developing brand identity and increasing brand awareness, brand positioning, values, and personality, and developing a promotion mix that would build trust and loyalty, and long-term commitment to community and patients.

5. Brand health tracking involves analysis of sales, market share, prescription, brand perceptions tracking, brand drivers, inhibitors, brand portfolio, investments, brand manager's capabilities, market, and competition.

6. A brand manager is primarily responsible for brand sales, growth, market share, profitability, and building brand value

7. The brand planning process involves situation analysis, SWOT analysis, brand strategy development, creating a single-minded proposition, developing marketing tactics, making brand Profit & Loss statement, monitoring & reviewing the execution, and developing a contingency plan.

References

1. Aaker, D. A., & Equity, M. B. (1991). Capitalizing on the Value of a Brand Name. New York, 28, 35-37.

2. Cobb-Walgren, C. J., Ruble, C. A., & Donthu, N. (1995). Brand equity, brand preference, and purchase intent. Journal of Advertising, 24(3), 25–40. doi:10.1080/00913367.1995.10673481

3. Leuthesser, L. (1988). Defining, Measuring, and Managing Brand Equity: A Conference Summary. Marketing Science Institute

4. Keller, K. L. (1993). Conceptualizing, measuring, and managing customer-based brand equity. Journal of Marketing, 57(1), 1–22. doi:10.1177/002224299305700101

5. PharmaVOICE Staff, Brand Building: Building the Brand in a Competitive and Digital Market, PharmaVOICE, March 2018.

CHAPTER **14**

INTEGRATED MARKETING COMMUNICATIONS

LEARNING OUTCOMES

After reading this chapter, you should be able to:

- Understand the concept of integrated marketing communications

- Gain knowledge about promotional aides in pharmaceutical marketing

- Learn how to go about designing creative campaigns

- Get an insight into the application of emerging technologies (AR, VR, IoT)

- Understand how to allocate a promotional budget

INTEGRATED MARKETING COMMUNICATIONS

Integrated Marketing Communications (IMC) represents one of the 4 P's of the marketing mix - Promotion. It encompasses Advertising, Personal selling, Sales promotion, Public relations/Events, Direct marketing, and Internet(online) marketing in combination to provide clarity, consistency, and maximum communication impact (Terence Shimp and J. Craig Andrews, 2013).

Three important aspects of Integrated Marketing Communications (IMC) are as follows-

- Consistent communication - one single message

- Effective communication - creates desired impact
- Efficient communication - create the desired impact in less time and less cost

The sole objective of IMC is to create a consistent and synergistic effect of various promotional activities and it is the main responsibility of the brand manager to ensure that.

Let's briefly understand each component of the promotion mix.

1. Advertising is paid form of non-personal communication of ideas, goods, and services in mass media by an identified sponsor in order to persuade.

2. Public relations focus on all the relationships an organization has with its various publics. By publics, we mean all the groups of people with which a company or organization interacts: employees, members, customers, local communities, shareholders, other institutions, and society at large. In essence, public relations is used to generate goodwill for an organization.

3. Direct Marketing creates a direct relationship between the customer and the business on an individual basis. Direct marketing is a system of marketing by which organizations communicate directly with target customers to generate a response and/or a transaction.

4. Personal Selling is an interpersonal influence process involving a seller's promotional presentation conducted on a person-to-person basis with the buyer.

5. Sales promotion refers to the provision of incentives to customers or to the distribution channel to stimulate demand for a product.

6. Internet marketing refers to marketing online which includes a website, social media marketing, search engine marketing, etc.

This chapter covers the pharmaceutical promotion mix comprising primarily of print promotional material. Advertising will be covered in the chapter on 'OTC marketing', Personal selling will be covered in the chapter on 'Salesforce effectiveness' & Internet marketing will be covered in the chapter on 'Role of digital marketing'.

PHARMACEUTICAL PRODUCT PROMOTION

The pharmaceutical industry traditionally relied heavily on personal selling - detailing of their products to doctors. Advertising of pharmaceutical products is restricted to medical journals, newsletters, visibility at clinics, hospitals, pharmacy outlets, etc. Advertising in mass media is done by pharmaceutical companies in case of OTC products or for corporate image building. Public relations/events are in order to build a corporate image or new product events through medical conferences, symposia, patient activation, etc. Direct marketing is in form of mailers sent to a database of doctors, emails, telemarketing, etc. Sales promotion is primarily in form of trade offers and incentives to the sales force. Internet marketing has caught up exponentially due to the pandemic and includes a whole gamut of promotional activities carried out online. The pandemic has forced the industry to adopt alternative modes of communication with doctors. There has been rapid digitization in the last two years of the pandemic. Given below is the figure (Fig 14.1) highlighting various promotional mix elements in the pharmaceutical industry (adapted from Vijay Bhangale, Journal of Medical Marketing, June 2008) & Gauri Chaudhari, The Perfect Pill, 2020).

PROMOTIONAL DECISIONS IN THE PHARMACEUTICAL INDUSTRY

- Identifying markets & audiences
- Determining objectives & tasks
- Preparing the promotional budget
- Selecting the promotional mix
- Evaluating & implementing the promotional strategies
- Feedback

PROMOTIONAL MIX ELEMENTS IN THE PHARMACEUTICAL INDUSTRY

- Personal selling: Medical Reps detailing the products to doctors with the help of print promotional material (described further in this chapter), samples, and gifts.
- Advertising: Specialized media like medical journals, souvenirs, seminars, medical symposia, mailers, etc.
- Sales Promotion: Trade offers (schemes), gifts to trade, gifts to doctors, incentives to salesforce
- Publicity: Organizing medical seminars, symposia, exhibitions, medical conferences e.g. APICON (Association of Physicians of India Conference)
- Direct Marketing: Direct mailers, email marketing, etc.

- Online(digital) marketing: Website, Social media marketing, Search engine optimization, Apps, Mobile marketing, etc. (as depicted in Fig 14.1)

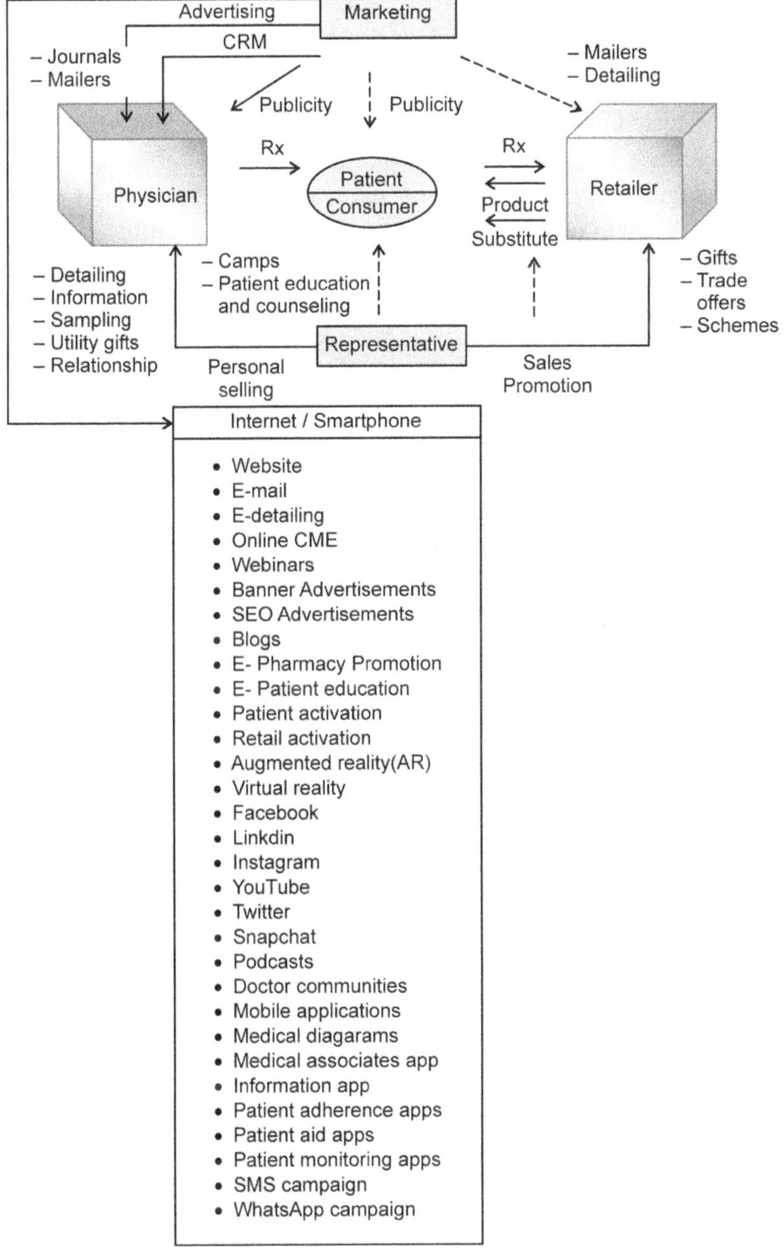

Fig. 14.1 Promotional mix elements in the pharmaceutical industry

Promotional inputs used by the pharmaceutical industry

Inputs can be broadly classified into two categories-

1. Print promotional inputs
2. Non-print promotional inputs

1. Print promotional inputs:

1. Visual Aid - this is a folder that includes dedicated page/s for various brand/s briefly highlighting the key brand messages, usage, packaging, etc., the main objective is to have a visual impact and aid the medical representative in detailing.

2. Product Monograph - this is detailed, comprehensive information on the product and the disease therapy. This promotional tool is normally designed at the time of a new product launch.

3. A product information brochure is a promotional input intended to focus on a specific indication or multiple indications.

4. Leave Behind Literature (also called LBL) is the input reinforcing the main message covering the information which could be of interest or use to a doctor. This input is left behind with the doctor so that it can be referred later, the intent is to serve as the Brand Reminder.

5. Reminder card, this is a small card with a simple message, which is normally left behind with the doctor to aid 'Brand Recall'. Also used in the situation where the doctor is busy and does not have time to listen to complete product detailing.

6. Teaser, this is a promotional input giving very superficial information about the new product, which is about to be launched. The objective is to generate interest among doctors and excitement among the field force.

7. Abstracts, articles, or re-prints, are normally the clinical trial reports substantiating the product claims or the therapy guidelines issued by a recognized medical body. The basic objective is to satisfy the customer's need for information.

8. Poster, this promotional input is used to educate the patient on a particular disease or therapy or to provide health tips that can facilitate the doctor's interaction with his patient. This also serves the purpose of a 'Brand Reminder'.

9. Frequently Used Questions (FAQs) are the most likely questions that the patients may have regarding a particular disease or condition or therapy and the answers to these questions are in simple language. The objective is to facilitate doctor's practice, also serves as the brand reminder.

10. Journal Advertisements are given in reputed medical journals to provide information and keep the brand on top of the doctor's mind. Inserts are the loose printed leaflets that are inserted in the journals.

How to design a creative campaign?

Visual-Aid or the Detailing Aid is the main communication tool used by the pharmaceutical industry for face-to-face promotion.

There are different copy formats currently used by pharma companies.

Whichever format is used, A copy should be a harmonious arrangement that facilitates smooth and fluent detailing of the brand's Unique Prescribing Propositions (UPPs)

- Strategic considerations could be
  - Whether launching a new product, in which case either it is a standalone communication or appearing at the beginning of the visual aid

  - Whether promoting an existing product, in this case, the emphasis given to the product will depend on how important the Brand is to an entire portfolio of the company's products. Accordingly, the placement and length of the copy will vary.

  - Visual aid sheet—to be shown and taken back, or an LBL, this would depend on whether there is a need for a brand reminder, maybe in case of a brand which is new and not yet in top-of-mind of a doctor or whether there is a need for information, e.g., a trial report which needs to be left behind for the doctor to go through later.

  - Counter competitor's claims (the 'knock-out' strategy), the objective here would be a specific task of taking a share from a particular brand, hence copy would highlight how the company's brand is superior over the competitor's brand, the comparison is done indirectly without taking the brand name of the competitor.

Four copy formats can be adopted for designing the medical copy.

Let's understand these four formats.

A. IEC / IEW format stands for

- Introduce

- Elaborate

- Conclude / Wrap-up

In this format, first a particular situation, typically a problem is introduced, e.g., 'in case of diabetic patients when the blood sugar levels do not normalize after treatment with regular anti-diabetic products or 'in case of urinary tract infections which do not respond to routine antibiotics.

The second step is to elaborate on how your brand i.e., the product will help solve the problem. E.g., in the above example, it could elaborate on how your product helps to normalize the blood sugar levels in all types of patients or elaboration on 'how your product treats all types of urinary tract infections

The third step is to conclude or wrap up by reinforcing the product's USP and requesting a doctor to prescribe your product in a specific situation.

B. ASK format, stands for

(a) Arrest attention

(b) Substantiate

(c) Knock out competition

In this format, the first task is to arrest the doctor's attention, this is by making some curiosity arousing statements, a complete page could be dedicated to this and the product could be introduced only on the subsequent page of the visual aid. It is very important to identify the right issue or a problem that the doctor is facing as the curiosity arousing statement needs to be relevant.

Once the curiosity is aroused, the next task is to substantiate how your product is useful in solving the given situation. References could be given here to substantiate your claims.

Finally, state how your product is superior to the competitor in solving the problem stated.

C. AIDA format stands for

- Attention

- Interest
- Desire
- Action

In this format, the first task is to state a situation or problem that would help in catching the attention of the doctor as seen in the earlier format.

Once the attention is grabbed, generate interest, e.g., give interesting data concerning the situation, then highlight the USPs of your product to generate desire, and finally close by reinforcing the product's USP and asking the doctor to prescribe your product in a particular given situation.

D. PAPA format stands for

(a) Promise

(b) Amplification

(c) Proof

(d) Action

In this format, you begin with a promise that your product is going to deliver, then amplify the benefit of this promise, give proof in form of trial reports, references, etc., and finally ask for a prescription.

It is important again to identify a specific situation, ensuring that your product is the best in treating that situation and having sufficient proof backing your amplified claims.

Visual Aid content in pharma could be

- General, all products included in one single visual aid folder, or
- Specialty-wise – different sections for General practitioners, Cardiologists, Gynaecologists, etc.
- E.g., a Cardiovascular division can have combined visual aid for
    - Cardiologist
    - Diabetologist
    - Nephrologists
- Frequency of Visual Aids could be: Quarterly/ 6 monthly/ yearly
- The flow Visual could be in the following sequence as follows:
- Brand name
- Composition, key ingredient/s in the brand

- The positioning statement is a key statement that will help you to create an image of your brand in the doctor's mind.
- Technology- Science, the backup data on the product, trial reports, etc.
- USPs, key benefits of the product which establishes the superiority of your product over competitors
- With Pay-off line, a tag line, which is unique, easy to remember, and captures the essence of how your brand is unique in solving the problem.
- Action (Demand/desire for Rx, purchase)

II. Non-Print promotional inputs: Non-print promotional inputs include

- Product samples to induce trials
- Doctor utility gifts (reminder items)
- Doctor CMEs, symposia, workshops, etc.
- Patient medical camps, free check-ups
- Trade offers and gifts
- Incentives to salesforce
- Internet marketing
- User trials(post-marketing surveillance studies)
- Developing patient groups for counseling
- Mobile marketing
- Developing apps

APPLICATION OF EMERGING TECHNOLOGIES (AR, VR, IoT)

Internet of Things (IoT) is transforming how healthcare is delivered. Health IoT constitutes any device that can collect health-related data from individuals, including computing devices, mobile phones, smart bands and wearables, digital medications, implantable surgical devices, or other portable devices. The benefit of digital for doctors is that it enables easily digestible information, quick responses to their queries, and interactions that fit with their schedule. Digital interaction with doctors needs to extend beyond the doctor-rep encounter and should involve a more tailored approach based on the behaviors, motivations, prescribing habits, and areas of specialty of physicians. Digital interactions might range from webinars to training interactions or videos, to online marketing, to physician-to-physician engagement **(PharmaVoice, May 2021).**

Augmented Reality (AR) is being used for educating patients regarding their medical condition and the action of medicines in their bodies. AR helps a patient to take a look at how a medicine works in his/her body instead of going through the long description put on the packaging. When a new drug is produced by a pharmaceutical company after extensive research, its usage needs to be showcased in an understandable way to the doctors, hospitals, regulatory bodies, and buyers. AR comes into play during this time. Salesforce training can also be simplified with AR-based tools.

Doctors are prescribing video games and virtual reality (VR) to treat conditions like attention deficit hyperactivity disorder (ADHD), depression, and post-traumatic stress disorder (PTSD) (Prescription gaming, JWT Intelligence, May 11, 2021)

Big-data, AI-driven engine is playing a significant role in pharmaceutical marketing. It can be used to profile a doctor, to know where he or she is on the product adoption continuum and which channels he or she prefers. It can also suggest appropriate personal and non-personal engagement activities that match that doctor's personalized profile, can guide as to what should be the next message and through which channel based on what has been done in the past and what has been successful with similar doctors in the same situation. (Amplity Health, 2020)

Technology is available to transcribe phone discussions and analyze voice patterns, keywords, and cadence. These data can be used to automate call reporting, identify training needs, and feed the database with information on call characteristics that impact prescribing. Data companies compile or generate the critical inputs for commercial strategies, analytical models, and more. Data sets can include HCP profiles, markets, social/media, EHR/claims/Rx, research publications, insurance and government, socioeconomic and behavioral, lab/ diagnostic, ERP, budgets, personnel/performance, channel engagement metrics, and more (Pharmaceutical Executive, Feb 2022)

ALLOCATION OF BUDGET

Budgeting is a critical part of IMC planning. Allocation of budget among the brands will depend on the contribution of the brand to the overall revenue of the company, brand growth rate, market growth rate, future prospects, level of competition, company stakes in a particular therapeutic area, etc.

The choice of promotion mix will depend on several factors like product life cycle, therapeutic area, geographic location, individual doctor choice, etc. Budgeting can be done taking into account the following aspects:

1. The objective and the task

    - The objective could be to increase market share by taking prescription share from a particular competitor or a particular doctor segment

    - Deciding the tasks be carried out in order to take prescription share from a particular competitor or a particular doctor segment

    - Estimating the expenditure for carrying out the above tasks

2. Setting budget as a fixed percentage of sales (last year's sale or anticipated sale for next year)

3. Budgeting via competitive parity – budgeting based on examining what competitors are doing.

Although setting a budget based on objective and task is the logical way of budgeting in practice all the above aspects are taken into account for the allocation of the budget.

SUMMARY

This chapter deals with the concept of integrated marketing communications, Pharmaceutical promotional mix, promotional aides used in pharmaceutical marketing, designing of creative campaigns, application of emerging technologies (AR, VR, IoT), and allocation of promotional budget. Key points are as follows:

1. Integrated Marketing Communications (IMC) represents one of the 4 P's of the marketing mix - Promotion. It encompasses Advertising, Personal selling, Sales promotion, Public relations/Events, Direct marketing, and Internet(online) marketing in combination to provide clarity, consistency, and maximum communication impact.

2. The pharmaceutical industry traditionally relied heavily on personal selling - detailing of their products to doctors. Advertising of pharmaceutical products is restricted to medical journals,

newsletters, visibility at clinics, hospitals, pharmacy outlets, etc. Advertising in mass media is done by pharmaceutical companies in case of OTC products or for corporate image building. Public relations/events are in order to build a corporate image or new product events through medical conferences, symposia, patient activation, etc. Direct marketing is in form of mailers sent to a database of doctors, emails, telemarketing, etc. Sales promotion is primarily in form of trade offers and incentives to the sales force.

3. Internet marketing has caught up exponentially due to the pandemic and includes a whole gamut of promotional activities carried out online. The pandemic has forced the industry to adopt alternative modes of communication with doctors.

4. Promotional inputs in the pharmaceutical industry can be broadly classified into two types – print and non-print

5. Print promotional inputs are Visual Aid (VA), Product Monograph, Product Information Brochure, Leave Behind Literature (LBL), Reminder Card, Abstracts, Articles, or Re-prints, Poster, Frequently Used Questions (FAQs), Journal Advertisements, etc.

6. Non-print promotional inputs include product samples, utility gifts (reminder items), CMEs, symposia, medical camps, trade offers and gifts, internet marketing, user trials, developing patient groups for counseling, apps, etc.

7. Visual-Aid or the Detailing Aid is the main communication tool used by the pharmaceutical industry for face-to-face promotion.

8. There are different copy formats used by pharmaceutical companies as follows-

- IEC / IEW format - Introduce, Elaborate, Conclude / Wrap-up
- ASK format - Arrest attention, Substantiate, Knock out competition
- AIDA format – Attention, Interest, Desire, Action
- PAPA format – Promise, Amplification, Proof, Action

9. Emerging technologies like AR, VR, IoT are being applied in marketing pharmaceutical products to doctors, for patient education and training of the sales force on new products.

10. Allocation of promotional budget depends on several factors like product life cycle, therapeutic area, geographic location, individual doctor choice, etc.

REFERENCES

1. Shimp Terence and Andrews Craig J., Advertising Promotion and Other Aspects of Integrated Marketing Communication, Boston: Cengage Learning, 2013

2. Bhangale Vijay, Pharma Marketing in India: Opportunities, Challenges and the Way Forward, Journal of Medical Marketing, Vol. 8, Issue 3, June 2008.

3. Gauri Chaudhari, The Perfect Pill, 2020, pp.251, Sage Publications India Pvt. Ltd.

4. COVID and the Rapid Rise of Connected Health, PharmaVoice, May 2021

5. Amplity's Blended Engagement Model A New Post-COVID 19 Health Care Provider (HCP) Customer Engagement Approach for Pharma, An Amplity Health White Paper, Oct 2020.

6. Swanson Randy, Connected Health: Covid and the Rise of Connected Health, PharmaVoice, May 2021

7. Prescription gaming by JWT Intelligence, May 11, 2021, https://intelligence.wundermanthompson.com/2021/05/prescription-gaming/ accessed on Oct 10, 2021.

8. Rawlinson Jonny, The Tech Perspective Coordinating a plan of engagement, Pharmaceutical Executive, Feb 2022, pg.22

C H A P T E R **15**

SALESFORCE EFFECTIVENESS – KEY TO PHARMACEUTICAL BRAND SUCCESS

LEARNING OUTCOMES

After reading this chapter, you should be able to:

- Understand pharmaceutical selling
- Understand the changing scenario in pharmaceutical sales force requirements
- Understand the concept of sales force effectiveness
- Get an insight into sales force effectiveness in the pharmaceutical industry
- Understand sales force automation

PHARMACEUTICAL SELLING

Personal selling is a crucial determinant factor of success in pharmaceutical marketing. Pharmaceutical selling is a highly specialized form of indirect marketing as the products and services are marketed to customers like physicians, who in turn advise/ recommend the products through their prescriptions to the end consumer i.e., patients. The product cannot be purchased by the patient without the prescription of a physician, except in the case of the OTC products.

The domestic pharmaceutical market is highly crowded with off-patent, branded generic pharmaceutical products, offered by many dominant players with their enormous capacities, and capabilities to develop, manufacture and market these formulations. Their sales forces comprise well-trained, skilled, experienced, and massively numbered medical representatives (MRs) and line managers, spread across the geographies, which makes the industry intensely competitive. The branded generic products are characteristically 'undifferentiated' in terms of their compositions, prices, target customers, sales promotion and advertising, and at times, even strategies. Under the circumstances, in a fiercely competitive pharmaceutical market, increasing the size of the sales force gives a limited advantage which can quickly and easily be blunted with a counter initiative by other players. The competition is far less between products, strategies, and the size of the sales force. It is more about the competition between the talent of one company and another. It is the effectiveness of your sales force that gives a competitive advantage for the success of your company in the marketplace.

In pharmaceutical selling, the role of the sales force is indispensable, irreplaceable, and invaluable. The face-to-face interface continually between the physician and the company through the sales force (MRs) remains the most essential element of the pharmaceutical business as acknowledged by both sides. This is proven beyond any doubt over the years, despite some failed attempts and initiatives by pharmaceutical companies to permanently replace or eliminate the sales force, either due to circumstantial compulsions or for alternative experimentation.

Consequently, it is important to study the evolution of sales force structures over the last 3 decades as their role and function have undergone considerable changes.

Till the early 90s, pharmaceutical sales organizations were simple monolithic structures. The era of a multi-division, multi-therapy, specialty-focused approach was yet to happen.

Most large-sized, leading companies operating with hundreds of MRs spread across urban and rural markets, promoting all the brands through a single large sales force. The operational emphasis was on reach, frequency, and coverage while the sales force skill requirements were limited to effective detailing, selling ability, and good communication.

The line managers primarily functioned as 'supervisors', overseeing whether the MRs complied with the fieldwork norms or not.

With globalization, the Indian market witnessed the proliferation of more products, therapies, number of players, and growth opportunities in the acute & chronic therapy segments at a brisk pace, with companies struggling to increase market share, prescription share, and higher productivity per MR. Changing global economic forces, intense competition, rising costs, and evolving industry policies posed new challenges and threw up new opportunities for pharmaceutical companies.

The pharma companies soon changed their sales force structures to suit the specialty-focused multi-therapy, multi-product approach. Specialty-focused Business Divisions became the new norm. Sun Pharma was one of the first companies to adopt this structure and emerge as a pioneer to be emulated by Intas, Torrent, and others. Today, almost every major pharma player, domestic or MNC, has leaner sales force structures extending the number of Divisions.

Table 15.1 Changing scenario in pharmaceutical sales force requirements

Factors	70s & 80s	Beyond the 90s
Skills and Knowledge requirements	Primarily Communication, Selling Skills & Technical knowledge	Customer Management skills, Customer Relations. New business needs arise in the pharmaceutical selling process with the role of the sales force changing with emphasis on customer services, networking skills, teamwork, and interdependence. Ability to impart scientific information & knowledge gives a competitive advantage.
Field force Reach and Penetration	Mostly in Metro and Urban markets	Penetration into Class I/IA and Rural markets. Hospitals (private & public) opening up new potential.
Nature of competition	Low intensity, Few players	The intense competitive rivalry among players.
Disease Pattern	Acute	Chronic therapies like Cardio-vascular, Neuro-psychiatry, Gastro, Diabetes, Derma-Cosmetology, Ophthalmology, etc.
Product Range	Simple, Primary, Conventional, Low priced	Exhaustive, Modern, Innovative, Expensive. Biologics, Biosimilars, Preventive Medicines, Vaccines. Innovative drugs through in-licensing arrangements. Technology brings novelty to current molecules.

Contd...

Factors	70s & 80s	Beyond the 90s
Marketing Approach	Product-centric, information-sharing	Customer-centric, Relationship focused. Developing patient care and compliance management programs. More launches of patented products, at premium prices.
		Marketing efforts beyond doctors as an integrated healthcare system comprising health centers, diagnostic facilities, medical insurance & healthcare NGOs, and health workers will play a larger role.
Treatment Approach	Therapies targeted at diseases	Preventive, Maintenance, Disease Management comprising diagnosis, treatment, monitoring, lifestyle changes, dietary controls & longer 'patient-product' association.
Salesforce structures	Large-sized Salesforce, Single SBU.	Multi-divisions SBUs. Spread across specialties. Less number of Salesforce per SBU.

SALES PRODUCTIVITY & SUSTAINABILITY

From the sales point of view, increasing the Per Capita Per month (PCPM) i.e. the Yield per MR per Month, a key performance measurement criterion, is an outcome of 4 variables i.e. Knowledge, Activity, Skills, and Motivation of the MR. Any shortfall in one of the variables impacts the performance outcomes. The line managers' focus, therefore, moved from a 'Simple Supervision' to 'Effective Intervention' and 'Direct Contribution' in these 4 key result areas towards Improving the Productivity of the MR.

INTRODUCTION TO SALES FORCE EFFECTIVENESS

The sales force is perhaps the most highly empowered organization within many companies. Usually working alone and unsupervised, salespeople are entrusted with a company's most important asset—its customers. Because of the sales force's critical impact on customer relationships, its effect on top-line performance is significant. Sales leaders agree that every sales force has the opportunity to improve sales revenues through enhanced Sales Force Effectiveness. Most companies realize the importance of maximizing the effectiveness of their sales organizations.

Despite a strong level of corporate attention, the concept of "Sales Force Effectiveness" is having different meanings for different people. A vice president of sales might view effectiveness as "adding value to the customer beyond the product by changing the sales process from transactional to consultative." A sales compensation analyst might view it as "increasing sales force morale and motivation through better incentive compensation programs." A sales training manager might view effectiveness as "increasing salesperson competency through innovative training programs." A finance manager might view it as "increasing sales per salesperson" or "holding sales force costs below a benchmark percentage of sales."(Andris A. Zoltners et., al., 2008)

SFE & Pharmaceutical Industry

Pharmaceutical Sales Force Effectiveness (SFE) is a strategy put in place to enhance the impact of a pharmaceutical company's sales effort. SFE processes enable MRs and field managers to prioritize their work by providing them with important information, direction, and recommendations. Rather than leaving everything to the competence of the individual line managers, which used to be the case in the earlier era, today pharma companies recognize that Sales Force Effectiveness is a multi-nodal function. While Sales Performance Management falls under the accountability of line managers, it is not anymore confined, or restricted to them; rather SFE integrates sales management, marketing management, marketing research, distribution, and customer relationship management for ensuring performance and success, thereby; strengthening the line management fully. Thus, SFE gives rise to an organized, strategic, and operational system to bring efficiency, effectiveness, and sustainable competitive advantage to the sales force.

However, there is no silver bullet with which pharmaceutical companies can increase Sales Force Effectiveness. So, what drives sales force effectiveness? Undoubtedly, there is no simple answer. Salesforce Effectiveness consists of a set of initiatives woven around the business goals, to achieve excellence in sales performance and productivity enhancement.

The components under the domain of SFE in a pharmaceutical company are:

1. **Choice of customers and market segments:** The right customers and the market segments are critical for the company as huge resources go behind the selling effort. Most course corrections take place at this point based on the type of customers, the number of targeted customers and their geographic spread, the intensity of competition, and the product mix.

2. **Salesforce size (number):** This is a crucial determinant that takes into account the strategic reach & penetration at the customer level on one hand and the competitive forces on the other. Often, the

competence of your sales force drives your sales force size while their strategic deployment provides a competitive advantage.

3. **Span of control:** Today the span of control has reduced considerably as the front-line managers perform multiple tasks from supervising to supporting to substantiating the field effort. SFE enables and promotes effective controls.

4. **Territory Design/Layout:** This optimizes the cost, time, and physical effort required to cover markets, and customers thereby controlling field and operational expenses commensurate with the yield. The territory knowledge plays a vital role in this.

5. **Coverage & Frequency of calls (i.e. visits) to HCPs:** The crux of pharmaceutical selling is consistent & persistent coverage and frequency of listed doctors, chemists, number of calls per day (call average),

6. Importantly the 'Missed Calls' can be costly lapses resulting in loss of prescriptions, prescribers, and performance. It simply amounts to a case of mismanagement of territory operations which no company can afford, but most companies fail to be cognizant of. It is believed that just the regular, disciplined coverage activity by itself can sustain a territory's business for a consistent period.

7. **Input utilization:** Effective utilization of promotional inputs is considered to be of vital importance. The inputs include many elements like samples, gifts, print materials, information, and detailing aids. Besides the MR's skills, a lot depends on precise and timely utilization to ensure the desired response in terms of prescription generation or indents. There is also a darker side to it as promotional inputs are vulnerable to being under-utilized or not utilized or get misused. SFE provides a vital role in monitoring the utilization commensurate with the expected outcome.

8. **Recruitment, induction & orientation methods:** Getting the right people, inducting them through a well-defined program, and ensuring that the new person becomes productive at the earliest is critical to the performance. SFE enables these processes.

9. **Imparting knowledge & skills:** Training & development both impact performance. The SFE integrates by identifying needs, organizing classroom & OJT (On-the-Job) training, and updating and up-grading knowledge, and information levels of the salesforce, thus enhancing the competence of the salesforce. Digitization has led to the generation of huge customer-related data. Leveraging these data

insights for personalized communication with a healthcare professional is an important skill that the sales force needs to poses.

10. **Effective supervision, evaluation & control:** SFE as a system collects collates and conveys all the vital information related to the market, competition, and field activity of the salesforce thereby facilitating prompt, proactive, and precise decision-making by line managers across the hierarchies

11. **Intervention & contribution by line managers:** It is critical in sales operations for line managers to be vigilant, alert, and radar-like picking up 'Early Warning Signals' from reports, observations deviations, activity analysis, correlating with territorial output to take corrective steps. What is most desirable is that the managerial intervention should happen promptly to enhance the performance. Also, the line managers are required to contribute their bit of share in generating business during territory visits, and joint fieldwork besides addressing operational bottlenecks which hamper performance. SFE strongly focuses on both these aspects.

12. **Sales performance management:** Today, all decisions, actions, evaluations, assessments, and course corrections are information-based, and data-driven thanks to SFE and its systems.

13. **Motivation, rewards, recognition:** While motivation, rewards, and recognition have been recognized with due importance in managing performance, a systematic approach for keeping sales forces continuously performing is enabled with SFE.

Interestingly, though all factors covered above are relevant and essential in every company, their significance will bear not the same or equal importance at all times. It is driven by challenges, opportunities, and strategic marketing objectives prevalent in the company from time to time. Nevertheless, Salesforce Effectiveness has become a formidable, standard requirement in every company; especially when the pharmaceutical companies are operating with large sales forces, multi-product, multi-therapy, multi-division operations which beyond the control of an individual's experience, expertise and competencies. Thus, the industry has to come to heavily rely on Sales Force Automation (SFA).

SALESFORCE AUTOMATION

There are innumerable types and packages customized for the companies considering their sales structures, business models, and the size of the field force. However, no system is a solution for sales performance; but a tool to facilitate sales performance. When companies adopt SFA, irrespective of their product features, this hard truth cannot be ignored. These six facts that need to be blended with any SFA system are as follows:

1. Any SFA is not a cure for all. It is not designed to be. It is just a tool that needs to be used most appropriately while taking cognizance of factors relevant to your business, sales model, challenges, and objectives. Improper usage may completely disappoint the users regardless of the promise or expectations.

2. Involving, engaging the sales force and convincing them about its requirement, usefulness, advantages and operational ease is of prime importance to establishing SFA in the organization.

3. The commitment to use SFA by the sales force and the controlling authority at company headquarters may not be aligned and hence may give rise to dissatisfaction with the system at either end or both. So, a sound controlling mechanism to avoid such gaps must be in place.

4. Expecting too much; too soon is another problem among the SFA users. Managements raise the expectations of the sales force too high to gain quick acceptance and in the process, the expectations from SFA grow manifold to rapidly result in disappointment. Therefore, caution and patience are required to get accustomed to the SFA.

5. There is a strong element of control with the management over the sales force that SFA brings in. The salesforce at times might get into the pitfalls of conflicts or competition with the management at the cost of full returns of the SFA. The management should therefore ensure that competition is replaced by cooptation.

6. In the final analysis, there is a critical distinction between management encouraging technology use within the sales organization and management administering a successful technology initiative that improves sales performance. It requires the discipline to disentangle worthwhile oversight from excessive monitoring and the wisdom to understand that cooperation and competition must coexist. In other words, it requires 'managing'.

SUMMARY

This chapter deals with an understanding of pharmaceutical selling, changing scenarios of pharmaceutical sales force requirements, sales force effectiveness in the pharmaceutical industry, and sales automation. Key highlights are as follows:

1. Pharmaceutical selling is a highly specialized form of indirect marketing as the products and services are marketed to customers like physicians, who in turn advise/ recommend the products through their prescriptions to the end consumer i.e. patients.

2. The pharmaceutical sales force structure has undergone a sea change in the last three decades. From a simple monolithic structure in the early 90s to structures today are leaner, multi-divisional focusing on multiple therapies with multiple products.

3. Pharmaceutical Sales Force Effectiveness (SFE) is a strategy put in place to enhance the impact of a pharmaceutical company's sales effort. SFE processes enable MRs and field managers to prioritize their work by providing them with important information, direction, and recommendations.

4. Different components under pharmaceutical SFE are i) choice of customers and market segments ii) size of sales force iii) span of control iv) territory design and layout v) coverage and frequency of visits to doctors vi) input utilization vii) recruitment, induction and orientation methods viii) imparting knowledge and skills ix) effective supervision, evaluation, and control x) intervention and contribution by line managers xi) sales performance management xii) motivation, rewards, and recognition.

5. SFA is a tool that needs to be used most appropriately while taking cognizance of factors relevant to your business, sales model, challenges, and objectives.

REFERENCES

1. Zoltners Andris A., Sinha Prabhakant, and Lorimer Sally E., SALESFORCE EFFECTIVENESS: A FRAMEWORK FOR RESEARCHERS AND PRACTITIONERS, Journal of Personal Selling & Sales Management, vol. XXVIII, no. 2 (spring 2008), pp. 115–131.

C H A P T E R **16**

RELATIONSHIP MARKETING

LEARNING OUTCOMES

After reading this chapter, you should be able to:

- Understand that building relationship is the key to success
- Get an insight into engaging with healthcare professionals
- Understand strategies for building relationships
- Learn about sustaining and nurturing strategies
- Learn about managing Key Opinion Leaders (KOLs)

BUILDING RELATIONSHIPS – KEY TO SUCCESS

Corporates are focussing on building sustainable, competitive advantages by developing and maintaining close, collaborative relationships with a limited set of customers. Through these relationships, firms create value by differentiating their offerings. The attractiveness of this approach for building competitive advantage has led numerous marketing consultants to suggest that relationship marketing - the focus of marketing activities on establishing, developing, and maintaining collaborative, long-term connections is the new marketing paradigm.

Relationship Marketing aids you in comprehending your customer through its offline and online behaviour. It helps you in increasing touchpoints with the customer and accompanying them in their journey to purchase a product or service. Advances in technology have helped in better dividing the market territories, enhancing communications with customers, and providing an environment rich with information contribute to improving efficient strategies to affect customers.

The key role of marketing is to spread awareness, and attract and satisfy the customers, while relationship marketing helps retain the profitable customers and enhance revenue from them. In brief, marketing acquires the customers whereas relationship marketing retains the acquired profitable customers and enhances profitable business from them. Thus, relationship marketing is a long-term proposition to create a loyal and profitable customer base.

ENGAGING WITH HEALTHCARE PROFESSIONALS (HCPs)

In maintaining relationships with HCPs, Medical Representatives play a vital role. Medical Representatives assist in the formation of long-term relationships with HCPs. As the primary link between the company and HCP, they have considerable influence on the HCP's perceptions about the company's reliability and the value of the company's services and consequently the company's interest in continuing the relationship. HCPs often have greater loyalty to Medical Representatives than they have to the company's employing the salespeople.

Relationship marketing is viewed as the ongoing process of engaging in cooperative activities and programs with intermediate and end-user customers to create or enhance mutual economic value at a reduced cost. In the pharmaceutical industry companies develop relationship-building programs with physicians so that in times of increased competition, they can successfully retain their current customers. Today's HCPs have choices and with the growing digital presence, HCPs have become more aware, vocal, and powerful. They switch more often, hence retention has become challenging. We all know that retaining the customer is more effective than recruiting a new customer. The main purpose of relationship marketing is to win customers and make them profitable over time. (G. Pandit Pathak and S. Shankar Bhola, 2014).

For managing the relationship with HCPs and turning them into a profitable customer, the important parameter HCPs look after is the value provided by the company. Value is considered to be an important constituent of relationship marketing and the ability of a company to provide superior value to its customers is regarded as one of the most successful competitive strategies. This ability has become a means of differentiation and a key to the riddle of how to find a sustainable competitive advantage. By adding more value to the core product companies try to improve customer satisfaction so that the bonds are strengthened and customer loyalty thereby achieved.

Acting as relationship managers rather than merely messengers, a well-trained salesforce is encouraged to unleash their human sensitivity and form real and honest bonds with this principal group of customers. Therefore, there are several ways in which pharmaceutical companies can rebuild the kind of relationships that will yield the best outcome for all stakeholders in the equation. While human contact is certainly valued, when approaching physicians, less is most certainly more. Physicians want to choose how they were contacted, that they want contact with one or two representatives per company only, and that these representatives should be more responsive to doctors' needs. Relationships should be deeper, based on a clear exchange of objective and neutral scientific information, and finally, unbiased by commercial arguments.

STRATEGIES IN BUILDING RELATIONSHIPS

Relationship marketing used to mean developing a relationship between a sales rep and a physician, but today it represents so much more. From fostering relationships between patients, HCPs, industry, and advocacy organizations to cultivating a level of compassion and understanding that results in relationships with patients and patient communities, today a relationship marketing strategy has evolved into an approach that engages patients and physicians, and other stakeholders from the very beginning of any healthcare journey to the end (Pharma Voice, Oct 2019). Although pharmaceutical companies influence the prescribing behaviour of HCPs by engaging with them through personal engagement, nonpersonal engagement activities may be a more important driver than any personal engagement activities, highlighting the importance of high-quality clinical and RWE (Real-world Evidence) research, timely publication of key results and product-related information, and promotion of communication and experience sharing between prescribers (R. Jandhyala, 2020).

Due to Covid 19 pandemic, digital engagement with HCPs is increasing at a tremendous rate. Going forward, there will be a much larger variation in how doctors engage with pharma – from all face-to-face to all remote, to everything in between. As the industry adapts to the world that has been significantly impacted by the pandemic, the hybrid approach — where former field-based colleagues are doing both face-to-face and virtual or remote interactions — is fast becoming the optimal model for successful HCP engagement (www.iqvia.com).

Digital Engagement with HCPs includes analyzing online behaviour and indulging more touchpoints through banner and search ads. Pharmaceutical companies predominantly use webinars for medical education, and to engage HCPs in research or treatment discussion. They are also great for connecting HCPs with KOLs more frequently. But nowadays, due to increased screen time HCPs are moving towards a hybrid model of engagement with offline and online modes.

Engagement with HCPs must operate across several different channels to ensure that every engagement is on doctors' terms. Multichannel Marketing should be brought into the picture to escalate the engagement, which would involve offline as well as online channels and personal as well as non-personal. Pharmaceutical marketers, however, have always operated in a multichannel environment that included personal selling, direct mail, journal advertising, meetings, and conferences even before the internet era. Multichannel means getting synergies from connected channels.

The hybrid type of engagement model offers three significant advantages over the traditional ones to HCPs

(a) Relevant Communication - Flexibility of consuming content in different ways that best suit the HCPs.

(b) Better Clinician Education - Enables the HCPs to benefit from 'on-demand' educational content and guided e-learning tools.

(c) Improves Treatment Outcomes - Assisting patients through their healthcare journey, by assisting them in improving adherence to medications.

Pharmaceutical companies, too, can benefit significantly from the hybrid model:

(a) Increasing Marketing ROI: This allows you to measure the return on investment of your marketing materials and efforts. Because the channels are connected, you can also connect the data with a complete overview of what, when, and how content is being accessed.

(b) Empowers Field Force: Medical representatives can supplement their face-to-face meetings with physicians with online remote contacts such as virtual detailing, eMSL (e-Medical Sales Liaison) meetings, and other forms of virtual engagements. Besides, medical representatives can help physicians to access different channels with on-demand content and offer personalized content to suit their individual needs.

(c) Refining Segmentation: Many pharmaceutical companies are realizing and recognizing the need to refine their segmentation strategies by targeting high-priority customers with smaller field forces. The hybrid model enables a key account strategy by delivering customer experiences that match your refined market segmentation. (Subba Rao Chaganti, 2020).

Relationship Marketing helps businesses in successfully implementing strategies aimed at winning customers and gaining profit from them. It is also helping businesses shift from a short-term transaction-based mode of operation in their transactions with customers to a long-term relationship mode. The broad objectives are as follows-

1. To improve marketing productivity and to enhance mutual value for the parties involved in the relationship

2. To enhance marketing effectiveness by carefully selecting customers for their various programs, by individualizing and personalizing their market offerings to anticipate and serve the emerging needs of individual customers

3. To fulfill consumer's expectations and their goals related to efficiencies and effectiveness in their purchase and consumption behavior

4. To build customer loyalty and commitment and develop new products, and redefine the competitive playing field for the company

While developing a strategy for building relationships with the HCPs, the first and foremost thing to acknowledge is that not all customers give equal business to the company. An HCP portfolio comprises the mixture of groups that make up the base of a business, a particular HCP may give his/her preference to a molecule/brand. As it is righteously proven for pharma that acquiring a new customer is 5-25 times costlier than retaining an existing customer and 5% increase in retention rate, increase profits by 25% to 95%. The bottom line of each strategy is to retain and build a profitable customer base (S.Tripathi, 2020).

The key stages in building a relationship are as follows-

1. Building a foundation of trust — Without trust, no one will fully engage with a rep and truly be influenced by what he or she has to say. In every interaction, reps can learn to demonstrate that they have the best interest of the other person in mind. Delivering on commitments is a key component in building a foundation that will drive future success.

2. Developing mutual respect — Respect for the busy schedules of both the reps and the healthcare staff cuts both ways and requires that reps appreciate everyone's time is limited if they expect that courtesy to be extended to them. Respect also is created through deep knowledge regarding products, the industry, and key economic factors affecting doctors and pharmaceutical companies.

3. Creating rapport — The human element is an important variable in a relationship strategy. Everyone has different interests and preferred methods of interaction. Discovering and adhering to personal preferences demonstrates that sales reps care about the best interests of the people they work with.

4. Delivering value — Each player in a doctor's office will define success a bit differently. Reps who understand how people define success, and then position themselves, their products, and their services in a way that supports those goals, go a long way toward delivering the value that will keep people receptive to their messages, products, and services in the future.

SUSTAINING AND NURTURING STRATEGIES

Relationship marketing is being touted as an effective strategy to provide pharmaceutical companies with the competitive edge they need to succeed in the future. They are drawn to the idea of cultivating relationships in a market environment characterized by intense competition and by core product and service offerings which are rendered by technology to be similar, if not identical, to those of the competition.

A sustained competitive advantage exists when the value-creating strategy is currently not being implemented by an organization's competitors or potential competitors and when these other organizations are not able to imitate, either through duplication or substitution, the benefits of the value-creating strategy. For a strategy to be valuable means that it enables an organization to increase revenues by taking advantage of opportunities in the organization's environment, reduce costs by neutralizing threats in the organization's environment, or do both. The potential for competitive parity and normal performance may be decreased if the organization is not organized to exploit valuable resources and capabilities.

When a strategy is valuable and rare, it will be a source of temporary competitive advantage and may lead to above normal performance. This above-normal performance will persist until the strategy is imitated by enough competitors to create perfectly competitive conditions and reduce the temporary competitive advantage to competitive parity. The potential for temporary competitive advantage and above normal performance may be lessened if the organization is not organized to exploit resources and capabilities that are valuable and rare. For example, a pharmaceutical company marketing its own patented drug for 20 years and generating a huge load of money.

If a strategy is value-creating but not rare it may lead to competitive parity, but only if the firm is organized to fully exploit this competitive potential. Second, if a strategy is value-creating and rare it may lead to a temporary competitive advantage, but only if the firm is organized to fully exploit this competitive potential. Finally, if a strategy is value-creating, is rare, and is imperfectly imitable it may lead to a sustained competitive advantage, but only if the firm is organized to fully exploit this competitive potential.

In this new marketplace, HCPs are less inclined to maintain loyalties; they are seeking value from organizations, and are demanding that organizations provide a good reason for HCPs to deal with them. In HCP marketing, the result has been a greater emphasis on developing relationships with HCPs. This means less emphasis on the elements of the traditional marketing mix and much more on creating a situation where an HCP is satisfied with everything an organization has to offer, to the point where the HCP wants to do business with the organization well into the future. Increasingly, the attitude in customer-driven organizations is that the marketing mix is a given and that the successful organizations are those which can satisfy their customers, as much through how they treat them as through the products and prices that are offered and develop relationships with them. (Rowe G & Barnes J, 1997).

By using tools and technology such as predictive analysis, tracking activities, AI and consumer feedback, relationship marketing or CRM (customer relationship management) become a brand's road map to a multifaceted emotional connection with all of its consumers, which hopefully leads to retention and loyalty (Pharma Voice, Oct 2019). Relationship marketing refers to all of the varying elements necessary to facilitate such relationships. Personalization is one of the most important elements of effective relationship marketing, and marketers need to understand that every touchpoint is an opportunity to build brand affinity.

MANAGING KOLs

Key opinion leaders (KOLs) have always been around in the pharmaceutical industry. The same amount of emphasis is given to retaining and managing KOLs as that to HCPs. Effective KOLs are now one of the primary marketing tools. KOL management involves -

- **Engagement:** It is the predominant part of KOL operation and is designed around set objects the association has for each specific KOL or group of KOLs. KOL engagement conditioning may include participation in internal and premonitory board meetings and conferences, clinical study design, and KOL marketing material review. KOL relations and conditioning may be detailed in a formal engagement plan to ensure thickness and replicability. Tracking, auditing, and reporting KOL engagement is also essential to perfecting the overarching operation program.

- **Assessment:** Measuring the impact and success of a KOL operation plan is critical to enhancement and resource optimization. The performance and prosecution of agreed-upon conditioning should be covered using predefined crucial performance pointers. Other aspects of KOL operation include KOL mapping, which considers KOLs within the larger environment of the association's entire KOL portfolio; and KOL segmentation, which is the division of KOLs into groups grounded on their influence, collaboration preferences, and other characteristics.

When KOLs are successfully managed, nearly all stages of the pharmaceutical product lifecycle benefit from original medicine discovery to nonsupervisory and request entry stages. Conclusive medical and marketing material, speed up patient reclamation for trials, wide-reaching trial results, strong consumer mindfulness of medicine products, and impacting HCP defining behaviours are some of the measurable impacts of successfully managed KOLs.

Pharmaceutical companies are frequently under pressure to give the results of their expansive, and frequently expensive, KOL operation conditioning. With defined processes, palpable pretensions, structured information reporting, and advanced relationship operation technologies, the true impact of pharma KOL engagement can be determined and assessed (Jeffrey J. Meffert, 2009).

SUMMARY

This chapter deals with building relationships, engaging with healthcare professionals, strategies for building relationships, sustaining and nurturing strategies & managing Key Opinion Leaders (KOLs). Key highlights are as follows: -

1. Relationship marketing is a long-term proposition to create a loyal and profitable customer base.

2. Relationship marketing is viewed as the ongoing process of engaging in cooperative activities and programs with intermediate and end-user customers to create or enhance mutual economic value at a reduced cost. In the pharmaceutical industry companies develop relationship-building programs with physicians so that in times of increased competition, they can successfully retain their current customers.

3. From fostering relationships between patients, HCPs, industry, and advocacy organizations to cultivating a level of compassion and understanding that results in relationships with patients and patient communities, today a relationship marketing strategy has evolved into an approach that engages patients and physicians, and other stakeholders from the very beginning of any healthcare journey to the end

4. As the industry adapts to the world that has been significantly impacted by the pandemic, the hybrid approach — where former field-based colleagues are doing both face-to-face and virtual or remote interactions — is fast becoming the optimal model for successful HCP engagement.

5. Key stages in building profitable relationships are – (i) Building a foundation of trust, (ii) Developing mutual respect, (iii) Creating rapport & (iv) Delivering value

6. Using tools and technology such as predictive analysis, tracking activities, AI and consumer feedback, relationship marketing, or CRM (customer relationship management) become a brand's road map to a multifaceted emotional connection with all of its consumers, which hopefully leads to retention and loyalty. Relationship marketing refers to all of the varying elements necessary to facilitate such relationships.

7. Personalization is one of the most important elements of effective relationship marketing, and marketers need to understand that every touchpoint is an opportunity to build brand affinity.

8. KOL management involves effectively engaging the KOLs and measuring the impact and success of a KOL operation plan

REFERENCES

1. G. Pandit Pathak; S Shankar Bhola (2016). Customer Relationship Management Strategy and Business Outcomes – Perception by Medical Representatives. Journal of Educational and Management Studies,6(1),01-08.
2. Robin Robinson, Relationship Marketing, PharmaVOICE, Oct 2019.
3. R. Jandhyala (2020). Influence of Pharmaceutical Company Engagement Activities on the Decision to Prescribe: A Pilot Survey of UK Rare Disease Medicine Prescribers. Pharmaceutical Medicine. 34,127-134.
4. https://www.iqvia.com/blogs/2022/02/embrace-hybrid-hcp-engagement-or-risk-falling-behind accessed on April 25, 2022.
5. Chaganti Subba Rao, 2020, Digital Pharma Marketing Playbook, PharmaMed Press, pg. 247-253.
6. Tripathi Shailendra (2020), CRM in Pharmaceutical and Healthcare Marketing, Clever Fox Publishing, pg.14-15.
7. W. Glenn Rowe; James G. Barnes (1998). Relationship Marketing and Sustained Competitive Advantage. Journal of Market Focused Management. Journal of Market Focused Management. ,2(3), 281–297.
8. Jeffrey J. Meffert (2009). Key opinion leaders: where they come from and how that affects the drugs you prescribe. Wiley Periodicals, 22(3), 262–268.

CHAPTER 17

MEDICO MARKETING

LEARNING OUTCOMES

After reading this chapter, you should be able to:

- Understand the concept of medico marketing

- Gain knowledge about the role of the medical services department

- Learn about medico marketing tools

- Get an insight into the evolving role of medico marketing

MEDICO MARKETING

Medico-Marketing is an exchange of information between a drugmaker and the prescribing doctor, or with the patients. This exchange aims to describe the company's products and how they will benefit the end-user in the diagnosis or treatment of the health problem. It also involves communicating medical knowledge or information for pharmaceutical marketing.

Typical Pharma Organization has departments like Marketing, Sales & Distribution department, Medical Services department, Production department, R&D department, Quality Control department, Finance department, Human Resources department & Commercial department.

ROLE OF THE MEDICAL SERVICES DEPARTMENT

Typical Medical Services Department comprises of Medico-Marketing, Clinical Research, Regulatory Affairs & Information services.

Medico marketing is involved in New Product Launches, New Product Evaluation & Interaction with Healthcare Professionals.

Traditionally pharmaceutical industry focused on the sales of the products, and the role of the Medical Advisor was limited to the training sales force and supporting the marketing claims of the product.

External changes in the Indian pharmaceutical industry like regulatory conditions and safety surveillance have increased significantly in the last decade or so. To adapt to these changes, the medical advisor's function has undergone a transformational change.

Medico Marketing works closely with the marketing team for developing strategies for product development, but at the same time, works at arm's length from them in the medical evaluation, when evaluating the comparative risk-benefit ratio of the drug.

Medico-Marketing plays an important role in carrying out the following functions:

1. Provide medical information

   Medico marketing provides accurate, fair, and balanced product knowledge, mostly in response to medical queries from consumers and healthcare professionals.

2. Provide product training to sales force and product management team

3. Create a database of scientific information related to the product

4. Review the promotional & training material

   Medico marketing has to ensure that the promotional and training material is in line with the applicable regulations and guidelines.

5. Provide scientific support to Sales & Marketing which can be used in promotion of product to doctors, pharmacists & patients.

6. Remain updated with journals and other scientific literature, to gain competitive intelligence

7. Design scientific communication in the following 3 cases –

   - Provide updated, unbiased, accurate, and complete information to the healthcare practitioner

   - For compliance with the Indian Drugs and Cosmetics Act, company guidelines

   - To interact with physicians on investigator-initiated studies, research activities, capacity building, and pipelines in drug development.

Medico Marketing acts as a vital link between internal stakeholders like Marketing Team, Sales Team, Business Development, R&D, Clinical Research, Pharmacovigilance & External Stakeholders like Doctors, Hospital authorities, Regulatory authorities, Patient groups & Payers i.e., insurance companies.

ROLE OF MEDICAL SERVICES IN NEW PRODUCT LAUNCH

It is involved in pre-launch activities like compiling and providing detailed information on the product for making the Product Manual, preparing training material on disease, therapy, and product.

It is involved during the launch in conducting a training program for Sales Team concerning the disease, therapy & product. Also, maybe involved in addressing healthcare professionals whenever it is a part of launch activity.

Post-launch, Medico Marketing is involved in replying to queries from doctors and sales force, providing additional information based on the request from doctors, product management, or sales team.

MEDICO MARKETING TOOLS

Medical Information Services is a sub-dept of Medical Services, it supports Medical, Marketing, and Sales to continuously provide updates related to products, therapies & suggest new products.

Inputs provided by Medical Information Services are broadly in these 4 areas:

- Product related information on existing products and competitors
- Therapy-related new "happenings" and guidelines on existing products
- Product related information on new products in terms of efficacy, safety &
- Competitor related information concerning new products in terms of superiority

Following are the medico marketing tools: -

PRINT PROMOTIONAL INPUTS

It helps in reinforcing the promotional theme and building brand-benefit association.

For example, let's say a company is launching a new inhalation therapy in the management of asthma that is the first of its kind for the usage which indicates to the patient when the metered dose in the inhaler is over and needs replacement. The print promotional material could be designed to reinforce the promotional theme highlighting the brand benefit. The brand recall inputs like product information brochures, Leave Behind Literatures, Articles could be designed highlighting the benefit as mentioned in the above example thereby satisfying customer needs for scientific information. Also, frequently asked questions (FAQs) on how to use the inhaler could be prepared and given to doctors to be given to patients, thus enhancing the interface with customers and thereby their involvement with the company and the brand.

Similarly, communication could be designed highlighting the unique benefit of the brand and sent to the field force thereby boosting their morale.

The print promotional inputs as mentioned in the example above also help in differentiating your brand from the competitors thereby enhancing field force effectiveness and providing a competitive advantage.

PRODUCT MONOGRAPH

Product monograph consists of rich, comprehensive, scientific, research-based information given in form of a book. Since the information is backed by the references quoted from reputed journals, textbooks, sites, etc. the product monograph acts as a credible and authentic source of information regarding the product and the therapy. Product monograph helps in strengthening the brand positioning establishing it as a science-based brand and is greatly valued by academicians and specialists. Product monograph is suitable for launching research molecule of a company or a new innovative concept or a new indication or usage.

Product monograph serves the basic purpose of providing information and thereby educating the customers. Good, authentic information provided in the product monograph helps the company in scoring over the competition. It also helps in establishing the company as an authority for a particular therapy and thereby building ownership in that therapy segment.

A typical Product monograph provides background information with regards to disease, its global implications, implications concerning India, reports published on the same, and Management of the disease. A Product Monograph also gives an introduction to the product, specifying its chemical name, its chemical properties, molecular formula, molecular weight, and chemical structure.

In addition, the product monograph provides a detailed composition in terms of ingredients and quantity of each ingredient, nature, and specification of the container i.e., the product packaging, special precautions to be taken for storage, pharmacokinetics, i.e., the absorption, distribution, metabolism, and excretion of the drug. Further, the Product Monograph contains information on pharmacodynamics, in-vitro activity, information in form of tables, graphs & therapeutic indications.

Product monograph should provide indication–wise studies, for example, if an antibiotic can be used in respiratory infection, gastric infection, and urinary tract infection, provide trial reports supporting the efficacy of the antibiotic in each type of infection. The trial reports should provide the data on efficacy, safety in comparison with competitor drugs. The data should be in form of tables, graphs, charts that are visually appealing and easy to understand. There is also information related to dosage, administration, and recommendation.

The Product monograph elaborates on the safety profile of the drug in terms of its tolerability, adverse effects which are serious, uncommon, rare, very rare, and overdose of the drug.

Product monograph provides the guidelines and recommendations for the management of disease or disorder. More the references mentioned in the product monograph better it is.

The line 'For the use of Registered Medical Practitioner or Hospital or a Laboratory only' should be added on the front cover or back cover of the product monograph.

PRODUCT INFORMATION BROCHURE

This promotional input is based on the specific target segment identified and the positioning of the brand. This is generally intended to focus on specific indication, specialty, disease management & segment. For example, let's say that the company marketing Sitagliptin, an oral anti-diabetic drug identifies Diabetologists as the target segment & positions the brand in Type 2 uncontrolled diabetes as the oral anti-diabetic drug that provides substantial glucose lowering without compromise. The Product information brochure should substantiate the claim by giving trial reports that Sitagliptin significantly reduces HbA1c, FPG(fasting blood glucose), and PPG (postprandial plasma glucose) levels without hypoglycemia or weight gain, thereby helping in managing diabetes. The scientific content is designed to deliver with strategic intent and is theme-based as we saw in the example of Sitagliptin.

The product information brochure provides good reading material for the physician and therefore generally carries high retention value. This facilitates brand recall as well. A product information brochure is useful for expanding the usage of the drug or widening the user base. For example, an existing nutraceutical brand can undertake a clinical trial finding new treatment uses of certain micronutrients and accordingly expand their usage in specific diseases or disorders among the doctors managing such conditions.

Leave behind Literature (LBL)

This is the input reinforcing the main message covering the information which could be of interest or use to a doctor. This is a low-cost input normally used in the follow-up stage.

This input is left behind with the doctor so that it can be referred later, the intent is to serve as the brand reminder. This has a high brand recall value. This promotional input is good for running a campaign, that is sequential or serial promotion based on a certain theme.

This is easy to use and thus user-friendly to the field force.

It can also be an effective tool against the competition. E.g., in the case of a cough preparation containing natural herbal ingredients, it can highlight the side effect such as drowsiness caused by cough preparations containing drugs. Thus, a theme-based LBL series can be made on letting your patients 'Lead Active life, without drowsiness' while treating cough.

The LBL highlights the indication, Unique Selling Proposition or Unique Prescribable Proposition, and the pay-off line or tag line. For example, in the case of a nutraceutical supplement, USP could be the balanced nutritional formula that supplements food in case of patients feeling weak and fatigued and the tagline could be Nutritional Supplement for 'Alert mind, active body. A theme-based LBL series can be designed based on this tagline.

Abstracts, Articles or Re-prints

These are normally the clinical trial reports substantiating the medical claims or the therapy guidelines issued by a recognized medical body. The basic objective is to satisfy the customer's need for information. It could be in form of an abstract of a particular study or a booklet of abstracts and full-text reference series. This can act as a differentiator and addresses the competitive situation.

PATIENT INFORMATION LEAFLETS, BOOKLETS, OR POSTER

These promotional inputs are used to educate the patient on a particular disease or therapy or to provide health tips that can facilitate the doctor's interaction with his patient. This also serves the purpose of a 'Brand Reminder'. For Example, patient education on 'Swine Flu' posters could be put up in doctors' chamber or patient education booklet on 'Do's & Don'ts concerning diarrhea can be distributed to patients through primary health centers in rural areas.

A diet chart for diabetic patients could be given to physicians, which can facilitate his interaction with diabetic patients.

These promotional inputs help in establishing the image of the company in a particular therapy area. These inputs can be made in regional languages so that the message is understood by the masses. This helps in spreading awareness about a disease or disorder & treatment and management of the same. Especially useful in lifestyle disease management like management of hypertension, diabetes, etc.

These promotional inputs, when aesthetically done, also add to the beautification of the doctor's chamber or an outpatient department (OPD) clinic in the hospital set up.

It is important to manage the customer relationship by seeking feedback, following up with customers, and encouraging the spread of word-of-mouth and thereby confidence building among the customers.

FREQUENTLY USED QUESTIONS (FAQs)

These are the most likely questions that the patients may have concerning a particular disease or condition or therapy and the answers to these questions in simple language. This helps in establishing the image of the company in a particular therapy area. It also helps in increasing the effectiveness of the field force. FAQs are designed keeping in mind the service orientation and hence based on identifying the customer's mission and goal. The end objective is to facilitate doctor's practice. FAQS are extremely useful in the management of chronic disease segments like cardiology, diabetology, neuro-psychiatry, rheumatology, osteoporosis, allergies, etc. FAQS are also useful for a new product launch or in case of new indications.

JOURNAL ADVERTISEMENTS

This is a form of scientific promotion. The objective is to provide information, build an association of the Brand or the company with a particular therapy or specialty and keep the brand in top-of-doctor's mind. Journals are accordingly chosen to give selective, strategic, and focused exposure to the brand and the company and provide a competitive advantage. Inserts are the loose printed leaflets that are inserted in the journals.

MEDICAL COMMUNICATIONS TO DOCTORS & FIELD FORCE

These are the medical updates or news bulletins sent to doctors and the field force. The objective is to stay connected and build an image in a particular therapy area. Communication regarding new concepts, ideas can be sent to the field force and doctors faster either through email or a printed mailer.

NON - PRINT MEDICO MARKETING TOOLS

- Webinars, Video Conferences, Blogs, Chat, etc.
- Websites, internet marketing
- User trials or PMS i.e., Post Marketing Surveillance studies
- Developing patient groups for counseling
- Free medical checkups or camps
- Doctor group activities, maybe a workshop, etc.

Let's first look at conducting CMEs.

CMEs form an important part of medical education, objectives of which are as follows:

- To build KOLs (Key Opinion Leaders) infrastructure – global, national, regional and local through consulting, training and education. Medical practitioners need to update themselves frequently as radical changes are taking place in the practice of medicine with the advent of new technologies, changes in healthcare delivery, changing demographics, and patterns of diseases. CMEs are organized for acquiring skills and knowledge that new trends in health care demand, and to motivate doctors to improve their performance and adopt continuous learning.
- To publish and communicate key data and information

- To present scientific and clinical information to the target audience encouraging the improvement of patient outcomes. CMEs also act as a vehicle in plugging the gap in the knowledge that has supposed to have developed over the years between the doctors and the changing healthcare patterns.

- To use data, theory, and experience to broaden the message and strengthen the product and the company image.

There are different forms or platforms of CMEs. Common forms are Medical Symposium or Seminar, Clinical Meeting, Panel Discussion or Workshop. These are normally conducted in medical colleges or corporate hospitals or could be conducted in clubs or hotels, which are often sponsored by pharmaceutical companies.

Medical Symposium or Seminar

This is a formal meeting in which several specialists deliver short addresses on a topic or related topics. The advantage is a lot of information can be transferred in a short period, however, it's the only one-way exchange of information and interaction is least.

Panel Discussion

This involves many speakers and the chairperson must have the capacity to extract the maximum from panel members. Discussion is on a topic that could be a particular challenge faced by the doctors in treating a particular disease or disorder and experiences or views shared by the panel members.

Medical Seminars or panel discussions could be arranged by pharma companies at the time of new product launches, especially if it's a new molecule. International speakers are often invited to such seminars.

Workshop

Workshops are usually brief educational program for a relatively small group of participants that focuses on special techniques and skills required in a particular field. In this, there is regulated discussion and learning are through demonstration and practice sessions. Workshops are more educative, interactive, and impart practical experience, but can cater to only a small number of participants.

How to conduct CMEs

General guidelines are as follows: -

1. From the brand or therapy strategy point, foremost important point is to brand the meeting - give a good name to the campaign.

2. Decide the number of meetings to be conducted across India.

3. Plan the Topic well in advance. Generate interest in the CME by publicizing it through field force, journal advertisements, digital media, etc. Topic related to recent advances in therapy, for example, something like 'recent advances in the management of diabetic foot' could be chosen to generate interest among the target doctors. Sequencing of sub-topics needs to be done carefully. The sequencing for the topic 'Recent Advances in Management of Diabetic Foot' could be done as follows:

   - Speaker 1: Diabetic Foot: Going through the basics

   - Speaker 2: Advances in Diabetic Foot Management

   - & Speaker 3: Antibiotic Selection – Novel options

4. The three pillars of a CME are Audience, Chairperson & Speakers.

   Who would be the Audience? You need to first define your objective and accordingly define the target audience. Need to decide whether the audience would be a Family Physician or a Consulting Physician or a Super specialist or a mix of all three. Also need to decide whether there will be different specialties or a single specialty. This would depend on your brand indications and the respective doctor specialties treating those diseases or disorders.

5. Once the theme & audience is decided, let's look at other Pre-meeting activities

   - Selection of a Chairperson - Different criteria to be taken into account would be level of seniority in the field, knowledge of recent advances, and capacity to extract maximum information from the panel members.

   - Selection of Speakers - Criteria could be experience and knowledge on the topic, good communication skills, etc.

   - Chairperson needs to have a set of queries that would be raised during the panel discussion.

   - Prepare the queries, data, and medical slides of the speakers.

   - Decide the venue of the meeting. Is it going to be a hotel or a hall or medical college auditorium? Arrangement of audio-visual equipment like slide project, microphones, etc., and catering arrangement accordingly need to be done.

- Pre-meeting planning to be done at the Head Office involves - designing & printing of invitation cards, designing of display material, backdrop, banners, material for an audience like a folder, meeting agenda along with the brief introduction of chairperson and speakers, abstract of their talk, writing pad with a brand name, a pen with a brand name, brand literature, etc.

- Thank-you cards, to be sent after the meeting.

- Also, need to keep ready the welcome note to be delivered, brief on the company profile, and note on a vote of thanks.

- In addition to the preparation from Head Office, Field force should get the confirmation on following before the meeting:

- Number of doctors who would attend the meeting

- Number of doctors who would probably attend

- The number of doctors who would not be in the position to attend the meeting.

6. Reminders must be given one to two days before the meeting as doctors are very busy it is likely to skip out of their mind.

7. On the day of the meeting, there is the welcome address given by the senior person from the company briefly giving the agenda and the theme of the seminar. Then introducing the chairperson and inviting him or her to take over the proceedings of the meeting. Once the discussion is over, the vote of thanks again is to be proposed by the senior person from the company. Once the meeting is over, during lunch or dinner, leverage the opportunity of interacting with doctors, in which the field force should be actively involved

8. Post-meeting, most important is follow-up.

- Follow-up must be done immediately after the meeting. There should be personal visits to all doctors who attended the meeting, thanking them for attending the meeting and seeking their feedback.

- The discussions in the meeting could be utilized as a promotion tool.

- The meeting proceedings can be captured in a video and used as a scientific promotion tool for the doctors who could not attend the meeting.

Such meetings are common and used by many pharmaceutical companies for medical promotion. Therefore, to stand out, give a different outlook by bringing in some element of novelty in form of a question-answer session or puzzle or case presentation, or a film.

PATIENT EDUCATION

Patients are consumers and every consumer at some time will be a patient or a caregiver for a parent, child, family member, or a friend. And as consumers, we have access to more information via more channels than ever before. Strategies should be based on reaching patients when and where they are focused on healthcare, searching for solutions, and being motivated to take action.

Marketers can better educate patients and provide tools for more productive conversations with healthcare providers.

Patients are the end-users of medicines, information to patients is part of therapy, patient education material ensures patient compliance thereby leading to a better outcome and thus helping doctors indirectly. This in turn helps in building brand image and company image.

Information given in patient-education booklets or posters is part of therapy like information on diet, activity, exercise, precautions, Do's & Don'ts as in case of diabetes or patients suffering from high blood pressure.

Factors affecting patient compliance could be

- Doctor related, e.g., cost of treatment
- Patient-related. e.g., inconvenient dosage
- Disease-related, e.g., a chronic disease requiring long term treatment
- Therapy-related, e.g., side effects of the treatment
- Treatment failure, complications, relapse, or recurrence

What differentiates between medical copywriting for doctors & medical copywriting for patients is information about the disease in terms of causes of disease, symptoms of disease & complications of the disease that would important to a doctor. Similarly, information about the drug in terms of mode of action of drug, dosage, duration, side effect, precaution, and other specific prescribing information would be more important to a doctor. Whereas, non-pharmacological management like diet, exercise, family care, and infection spread control could be more important to the patient. Similarly, Do's & Don'ts or 'Myths & Facts' are more relevant to the patient.

How to design patient education material

- Patient education material can be in form of one-page or two-page or four-page literature or the form of a small booklet. It should be in a reader-friendly format, in simple language, and even in local regional language wherever necessary so that the purpose of making and distributing patient education material is served.

- Posters can be designed for patient education to spread awareness about the disease, ensure patient adherence to therapy and other important information about the disease.

- Newer options like social media, videos, CDs, or pen drives can be used for patient education. For example, one of the leading pharmaceutical players in the asthma segment, Cipla Ltd. is running a social media campaign #Berok Zindagi highlighting the campaign's core message, "Asthma Ke Liye, Inhalers Hain Sahi". Similarly, Atlas of the human body or charts or models of human parts can be provided to doctors, which can be used to explain disease conditions to patients.

- Website development providing content, contact information, or answering queries is another good way of patient education and engagement. For example, one pharma company marketing vaccines for the prevention of cervical cancer has developed a brand website that provides patient information, prescribing information, important safety information, and information for health care professionals. Patients can not only interact with the company through this website to get answers to their queries but can also sound out if they are having trouble paying for the vaccine.

- Companies are setting up telephonic support for answering patient queries, guiding experts, and also providing medications support. As seen earlier one pharma company marketing anti-diabetic drugs has initiated a telephone-based complementary diabetes management patient support. The patients who have been prescribed the company's anti-diabetic drug can enroll in this program. Counselors engage with patients over the phone to provide diabetes management education and guidance on diet, exercise, monitoring, medication adherence, and other lifestyle changes.

- There are certain diseases or conditions in which patient education is an integral part of brand promotion. These diseases or disorders are diabetes, hypertension, rheumatoid arthritis, allergies, osteoporosis, asthma, psychiatry, etc. also conditions like depression, migraine, malnutrition or anemia, epilepsy, acne, and malaria, patient education is an important part of brand promotion.

MEDICAL CAMPS

Medical camps are mass gatherings of patients brought together with the purpose of mass treatment. Medical camps are organized usually by health trusts, charitable societies, NGOs, clubs like Lions club, Rotary club, political parties, government bodies, or medical associations

Objectives of Medical camps could be:

- Mass contact with patients
- For the social cause or a charity
- For creating awareness about the disease, treatment, prevention, patient education
- For building reputation, goodwill of the company
- For expanding the sphere of medical practice

Salient features of medical camps are:

- Free health check-ups like free blood sugar check-ups or bone mineral density check-ups etc.
- Targeted at groups like children, elderly, pregnant women, government offices, office employees, banks, industrial employees, or residential societies
- The tests could be free or subsidized
- There could be the distribution of free medication
- Patient counseling by experts could be there
- Association with specific days like Women's Day, World Diabetes Day, Doctor's Day, World Heart Day, etc.

How do pharma companies benefit by associating with medical camps?

Associating with medical camps helps pharma companies as follows: -

- Building company image as it is seen as a service to the medical profession
- Establishing direct association with a cause, disease, and treatment
- Building relationships with doctors
- Building a doctor-disease-therapy connect

- Establishing direct patient contact which is useful in lifestyle disease management
- Leveraging the captive business opportunity as all attending the medical camp could be potential consumers for their products.
- Gaining a competitive edge-
  - Medical camps provide a unique platform for 3 key stakeholders - pharma companies, patients, and trade to come together in one place providing the following benefits:
    - Provides direct access to the patients, high relevance to chronic lifestyle disease management wherein pharma companies can participate in patient education and patient group activities. This also helps in tracking the patient.
    - Retail chemist and stockist participation ensure supply of medicines, special rates could also be applicable.
    - Medical camps help pharma companies in customer relationship management, building relationships with patients, consulting doctors, and trade

There are different types of medical camps as follows: -

- Detection camps like asthma detection camp, diabetes detection camp, hypertension detection camp, osteoporosis detection camp, stress detection camp, gynecological complications detection like PID (pelvic inflammatory disease) or PCOS (polycystic ovary syndrome), etc.
- Disease control camps like blood pressure check-up camp for hypertensives, blood sugar check-up camp for diabetics, rheumatoid arthritis or osteoarthritis check-up, malaria, dengue check-up, HIV & AIDS check-up, etc.
- Determining progression of disease camps like retinopathy in diabetics, peripheral vascular disease (PVD) in diabetics, left ventricular failure (LVF) in hypertensives, chronic renal failure (CRF) in case of renal dysfunction, BMD or bone mineral density detection camps in Osteoarthritis, etc.
- Immunization Camp, vaccination for Covid -19, Hepatitis C, etc.

Medical camps are normally carried out at doctors' clinics, hospitals, primary health care centers, schools, or a community hall.

Material required for carrying out a medical camp is as follows: -

- Posters and information booklet for patient education

- Patient register to take down the details of patients
- Telephone numbers
- Information to be sent to local doctors
- Information to be sent to local NGOs, clubs, or health centers

A typical sequence of activities in medical camps is as follows: -

- Patient registration
- Height-Weight check-up
- Pulse, blood pressure check-up
- ECG recording
- Physician consultation
- Patient education
- Medication sample

Here are the ten mantras for a successful medical camp.

1. Organize the camp on a holiday or a weekend so that a good number of people attend the camp.

2. If you are organizing a specialty camp for a particular disease, consider keeping a small fee because it will help to filter 'genuine patients' from the free camp buffs who grace all the camps in town with their presence.

3. Take the help of a local Red-Cross authority, the Rotary Club, or some other NGO to impart some credibility to the camp. Use their facility for the camp. It is also a good idea to use the facility of a smaller nursing home or a clinic for the event.

4. Make sure you carry a lot of medicine samples to be distributed free to the patients.

5. Always market a camp as a screening camp for the procedures which would be carried out at the hospital at a later date. Announce a discount for the 'selected' patients. The ideal thing would be to rope in the NGO or some local government body or politicians or businessmen or all of these to bear some percentage of expenses for the procedures you would carry out on 'selected' patients if they are poor.

6. Build a hype of the camp at least four days before the event. Make your camp unique. Don't make it sound like just another camp by some hospital. This can be done using myriad marketing tools like - a) Pamphlets - give it a catchy headline, use short sentences,

mention free medicines, include a picture in the pamphlet to catch attention. These pamphlets can be used as inserts in the newspaper b) A loudspeaker mounted on the top of a vehicle and a person announcing the event date and other particulars. An announcement could be innovative like in form of a poem or a song. This helps to gain the attention of people in rural areas, c) Banners are put at a decent height so that they are easily visible. Put these at places like a bus stand or a busy crossing.d) Online promotion through social media.

7. During the camp systematically manage the crowd. Make sure that people do not have to wait for long hours and are taken care of.

8. Ensure that you are running audiovisual clips in the area while the camp is on. Also, give brochures and other reading material to the patients generously.

9. Follow-up a camp with a press conference. Ensure that the local press and cable TV people are present.

10. At the end of the day if you get the names and addresses of a few legitimate patients and if the press people send in a detailed report on your camp, you have done your job.

THE EVOLVING ROLE OF MEDICO MARKETING

The COVID-19 pandemic forced the pharmaceutical industry to shift how they need to engage with their customers - doctors as well as patients. The industry experienced the first-hand value of staying connected through more remote (i.e., phone, web conference) and non-personal (i.e., email, website) engagement. Medico marketing has a very important role to play in these new channels of engagement. Secondly, medico marketing is expected to play an important role in improving the efficiency of drug development. Thirdly, medico marketing will play an important role in involvement with payers i.e., insurance companies and patient groups in solving their issues. Fourthly, medico marketing can play a role in reaching the physicians and helping them in the capacity building of healthcare. Fifth, medico marketing can get involved in health economics by assisting in developing pricing models, negotiating with payers on the 'real value' of medicines, or capacity building.

Lastly, medico marketing would play a very important role in finding innovative and new drug delivery systems (NDDS) for the existing products and in developing a possible rationale for fixed-dose combinations (FDCs).

By 2025, Medical Affairs will elevate performance across medical activities to optimize experiences and outcomes for patients and physicians (A vision for Medical Affairs in 2025, Mckinsey & Company report). Digitization and automation is going to paly an active role in patient engagement and in providing medical information that Healthacare Professionals (HCPs) can immediately access whenever needed.

SUMMARY

This chapter deals with the concept of medico marketing, the role of the medical services department, medico marketing tools, and evolving role of medico marketing. Key points are as follows:

1. Medico-Marketing is an exchange of information between a drugmaker and the prescribing doctor, or with the patients. This exchange aims to describe the company's products and how they will benefit the end-user in the diagnosis or treatment of the health problem. It also involves communicating medical knowledge or information for pharmaceutical marketing.

2. Medical Services Department comprises of Medico-Marketing, Clinical Research, Regulatory Affairs & Information services.

3. Medico marketing provides medical information, product training to the sales force and product management team, create a database of scientific information related to the product, review the promotional & training material, provide scientific support to sales & marketing which can be used in the promotion of product to doctors, pharmacists & patients, remain updated with journals and other scientific literature to gain competitive intelligence and design scientific communication

4. Medico marketing tools are of two types - print promotional inputs and non-print promotional inputs

5. Print promotional tools are product monographs, product information brochures, leave behind literature (LBL), abstracts, articles or re-prints, patient information leaflets, booklets, or posters, frequently used questions (FAQs), journal advertisements, or medical communication to doctors & field force.

6. Non-print promotional tools are CMEs, medical camps, developing patient groups for counseling, conducting user trials or PMS i.e., Post Marketing Surveillance studies, digital engagement with doctors and patients

7. Medico marketing will continue to play an important role with the evolving new channels of engagement with doctors and patients like remote (i.e., phone, web conference) and non-personal (i.e., email, website) engagement. Also, medico marketing will continue to play a crucial role in new drug development, healthcare capacity building, health economics, and pricing models.

8. By 2025, Medical Affairs will elevate performance across medical activities to optimize experiences and outcomes for patients and physicians. Digitization and automation is going to paly an active role in patient engagement and in providing medical information that healthcare professionals can immediately access whenever needed.

REFERENCES

1. A vision for Medical Affairs in 2025, Mckinsey & Company report available at https://www.mckinsey.com/~/media/mckinsey/industries /life%20sciences/our%20insights/a%20vision%20for%20medical%2 0affairs%20in%202025/a-vision-for-medicalaffairs-in-2025.pdf

CHAPTER 18

DIGITAL TRANSFORMATION IN PHARMACEUTICAL INDUSTRY

LEARNING OUTCOMES

After reading this chapter, you should be able to:

- Understand the digital ecosystem

- Gain knowledge about components of digital marketing

- Learn about digital strategies for engaging with Healthcare Professionals

- Learn about digital strategies for engaging with patients

- Learn about measuring the effectiveness of digital marketing

- Gain insight into the near future digital outlook

DIGITAL ECOSYSTEM IN THE PHARMACEUTICAL INDUSTRY

Digital technologies are becoming more accessible in today's day and age, the advancement in creating content, dissipating it, and then monitoring and evaluating the outcomes are now at our fingertips. As smart devices become more and more accessible and affordable, they promise an enormous benefit from digitization to every individual and healthcare provider. These tools and services now have the power to generate transparency, and shared prosperity, strengthen inclusion and inspire innovation. Today, our digital ecosystems comprise multiple stakeholders and are based on three major pillars as follows-

1. **Digital Infrastructure**

2. **Digital Data, Social and Security**

3. **Digital Transactions and the Economy**

1. **Digital Infrastructure:** This pillar in the ecosystem signifies the parameters of connectivity infrastructure, digital literacy, ease of access and data availability - with affordability. Emerging digital technologies like 5G, Optical Fiber connectivity, Broadband systems are key parts of this pillar.

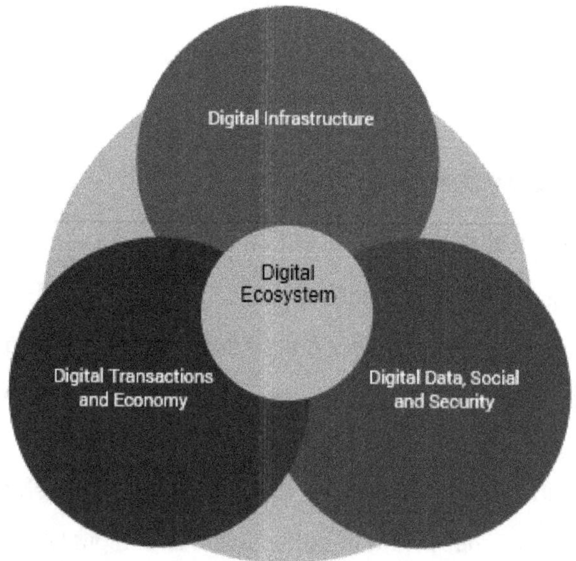

Fig. 18.1 Digital Ecosystem

2. **Digital Data, Social and Security:** A very definitive pillar in the ecosystem defines how data is analyzed and utilized, the social parameters form a basis of what content is served in context to the apt kind of user with data being secured and particularly intended for use by the individual or the company it is being served to, anonymity is a strong part of the reviews and ratings.

3. **Digital Transactions and the Economy:** This pillar explores the roles and opportunities for the adoption of digital financial services, e-commerce and e-pharmacies, digital trade, and real-time payment systems. This essentially complements the function of marketing, sales, and logistics. The pharmaceutical industry has adapted this pillar to a great extent with the inclusion of e-pharmacies and door

delivery systems in place for the patient or consumer directly and concerning the distribution system and retail pharmacies, the adaptation to e-transactions has been as early as 2016, with the introduction of the UPI (Unified Payments Interface) system. We will touch upon these pillars as we move in our journey to understand more about the digital transformation in the pharmaceutical industry.

Components of Digital Marketing

Different components of digital marketing are depicted in Fig 18.2.

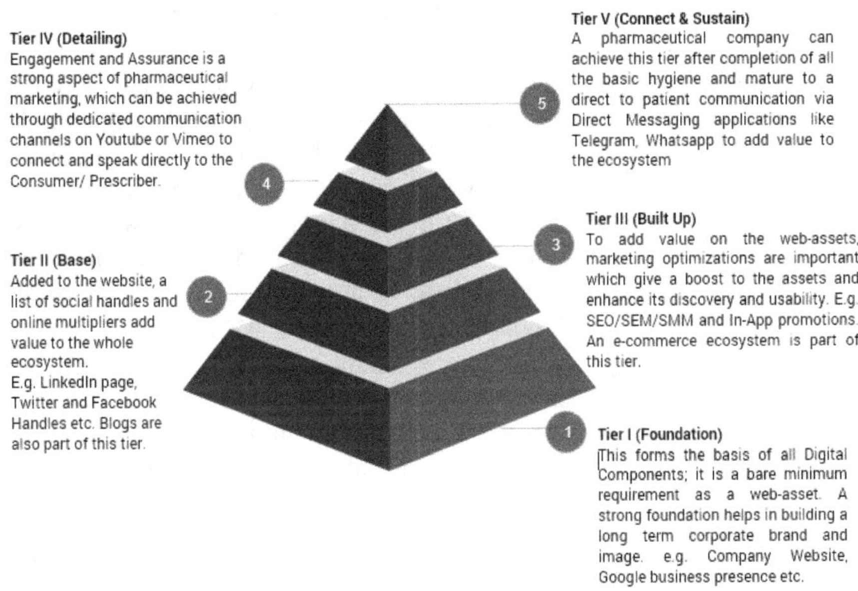

Fig. 18.2 Pyramid of the components of digital marketing in pharmaceutical industry

Let us briefly look at different components.

1. **Websites:** Websites being the most neglected web asset in pharmaceutical marketing, companies have been graduating to a more "web-first" approach in the post-pandemic scenario. The average time spent by a pharmaceutical company spent building a website is less than 60 hours, while other industries parallel invested more than 4X the time and resources getting their web presence right. *(Uplers Email - https://email.uplers.com/blog/email-marketing-for-the-pharma-industry/)*

With the ever-changing scenario in the post-pandemic culture, a clear path for outcome orient and value-added website systems is needed in the pharmaceutical industry. This will help assert the time lost over the years to come to digitization at a much quicker pace. A typical corporate pharmaceutical website is a dedicated space for stakeholders like doctors, distributors, and C&F partners, for Patient education, learning for a prospective employee, reporting adverse incidents related to drugs produced by the company, and for media consumption. This is the most important and self-owned web asset for a 24/7 presence, to have credibility, information exchange, and insights. A website is a direct company to stakeholder connection.

2. **Online Advertising:** Before the "hash tag" world, the most commonly used source for digital presence was display ads. The healthcare industry has been using them for the marketing of OTC products more often. Before the term D2C (Direct to Consumer) was coined we lived in a world of online-only brands which used online advertising in the form that involved bidding and buying relevant ad units on third-party sites, such as display ads on blogs, groups and forums, and other relevant websites. These types of ads include images, text, pop-ups, banners, and videos.

   Post this came to a whole new generation of marketing products viz. E-mailers, SEO, and SEM etc. Online advertising has currently become more content-driven and is maturing as we learn more about the innovations as they unfold. Retargeting is an important aspect of online pharmaceutical advertising specific to OTC products, which essentially requires a code that adds an anonymous browser cookie to track new visitors to your site. Then, as that visitor goes to other sites, you can serve them ads in various formats - display, text, AV for the products or services. This focuses your advertising efforts on people who have already shown interest in your company. As we speak the world is moving towards a cookie-free environment, the day is not far when this kind of advertising might become totally redundant.

   Pharmaceutical companies, except for a few exceptions, have been laggards in catching up with this kind of advertising. The catch here is most prescription products cannot be advertised in this mode due to strict guidelines and policies.

3. **Email Marketing:** Did you know that pre-pandemic, the average email open and click-through rates in the pharmaceutical industry were as low as 18.58% (Uplers Email - https://email.uplers.com /blog/email-marketing-for-the-pharma-industry/) and 2.25%

(https://www.aha.io/roadmapping/guide/marketing-plans/digital-marketing-plan-components) respectively?

By the mid-2020s the open rates were up by 27% and the click-through rates went up by 9%. To add to this worldwide research, email was one of the top modes of pharma-to-physician communications with 35 - 40 % of all forms of data being shared with the physician online. (*Uplers Email - https://email.uplers.com/blog/email-marketing-for-the-pharma-industry/*)

All these numbers clearly indicate that pharmaceutical email marketing works. Email marketing is a powerful medium for communication, information-sharing, and engagement, particularly if implemented as part of a multichannel marketing approach. But this requires marketers to focus on a few key strategies. Email marketing currently is used majorly for TOMAR (Top of mind awareness and recall), essentially for Information dissipation, e-commerce, and D2C orientation. This tool has regularly been used by new-generation pharmaceutical marketers for sharing the latest molecule guidelines and research studies with physicians and healthcare providers.

4. **Mobile Marketing:** From 2003 through to 2015, SMS marketing was an important component in the pharmaceutical industry; they were used by the CRMs (customer relationship managers) to share reminders for brand names, company details, and some updates of offers and promotions, for appointments with the physicians by the medical representatives and multiple other communication modes. The other important usage for SMS is to distributors and stockists for order booking and sharing information on stock availability and updates. Some companies use platforms with SMS-based reporting systems for pharmaceutical representatives and managers, which capture tower data showing the exact time of meeting with a physician or distribution channel.

Mobile marketing as a component has gained immense popularity in pharmaceutical in the recent year, due to these reasons:

1. Mobile device growth, options and availability

2. Mobile internet accessibility, affordability and network coverage

3. Innovation in applications, messaging, and optimization of content for mobile consumption

The experience of mobile phones for content consumption has increased drastically; this is due to the design focus on mobile-oriented websites, applications, and audio-visual content. Most pharmaceutical field executives do not have access to a personal laptop, even if they do, it is a difficult task to carry laptops daily to the field, a mobile devices have become more powerful than ever, which makes it easier to do more complicated computing tasks. The application ecosystem is another reason for pharmaceutical companies to have dedicated platforms to perform daily tasks by the employees.

5. **In-App Marketing:** Application ecosystems have developed well and are now more usable than ever. In a general scenario, mobile applications range from consuming & searching content, ordering and accessibility, information, and messaging to getting health reports and critical information.

In-app marketing is messaging, displaying content, and advertising to customers while they are actively using an application on their mobile devices. This gives businesses insights and data to share personalized and optimized content in real-time.

The OTC industry has evolved the use of this form of digital marketing over the years, with a click-to-action (CTA) attached to measure the yield of such a type of communication. The main analysis is the consumer's click-through rates and buying behaviors.

The personalization part is very important, which leads to showcasing of:

1. In-App Notifications

2. Splash Screens

3. Contests

4. In-App messaging for promotions and launches

5. Free trials for paid services

This is an ever-changing and evolving component and the growth opportunities for the brand to work with e-pharmacies and e-commerce is immense.

6. **Influencer Marketing:** The latest buzzword for most of us has been an influencer; it has replaced celebrity endorsement to a very large extent. The pharmaceutical industry has been using influencers for many years, as key opinion leaders (KOLs). The KOLs have a huge following and act as influencers to market certain products, and talk about indications and research.

Continuing Medical Education (CME) is a strong component of influencer marketing in pharmaceutical industry.

From being a strong offline-oriented content in the earlier years, it has graduated to platforms like Youtube, Facebook, and Linkedin. The organizations of today, in the post-pandemic world, are creating CMEs and Conferences online with a greater reach and wider audiences.

Podcasts and Channels on healthcare by the KOLs as Medical Thought Leaders and Influencers are gaining popularity; this has created an opportunity for marketers to be part of the journey for educating and promoting products and services. The evolution is far from over, influencers are destined to add value to our healthcare ecosystems with different specializations and innovative content additions.

7. **Retargeting & Remarketing:** Both of these terms have a different meaning, Retargeting mainly focuses on paid media to re-engage audiences who might have visited your website or social media handles, while Remarketing is a function to engage with customers who have already been doing business with your brand via emailers, messaging and other points of contact.

    The easiest way to gain attention and TOMAR is with relevant data, offers, and product information. Multiple tools and tricks are available to achieve this. Most of the retargeting happens via Search Engine Marketing (SEM), a topic which we will cover ahead.

8. **Search Engine Optimization & Marketing:** Did you know that there are more than 20 active search engines across the internet that can yield great results and are used by millions of people across the globe? The most popular search engines in India are Google, Bing, Yahoo, Rediff, and Duckduckgo. While by market share Google clearly seems to have a lead, a lot of searches are happening over voice-based engines like Alexa, Siri, and Youtube.

    Ailment searches top the list of healthcare searches, and medical conditions and reactions are a strong part of this scenario. The pharmaceutical industry has seen a huge spurt in post-pandemic search matrices for medication, conditions, and wellness products. The one major difference between Optimization and Marketing is that the latter is paid and the other is determined by skill and process, both these parameters yield amazing long-term results. The pharmaceutical industry has been ignoring these parameters for a long time, as the focus has always been on in-person meetings and promotions, the rapid adaption of SEO can be a marketer's strong strategy to set the company higher in the search ladder.

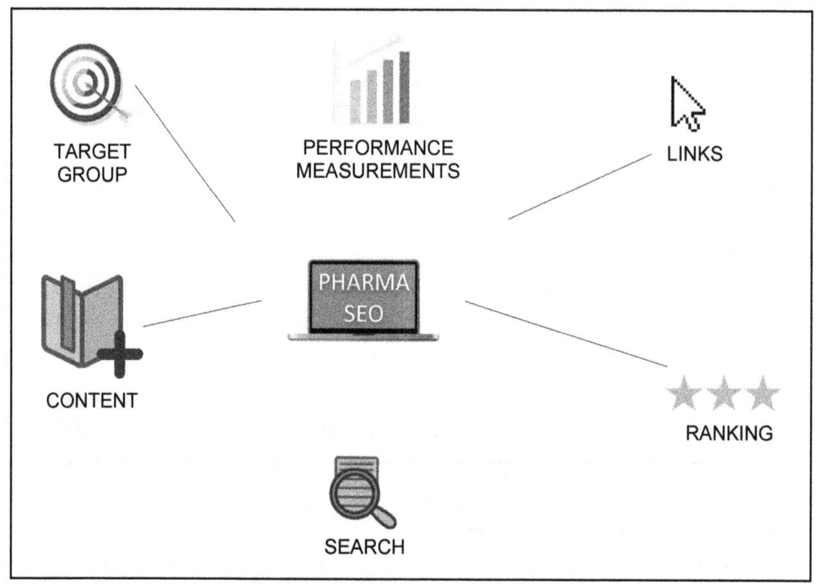

Fig. 18.3 Key aspects of SEO

Basic 3 fundamental principles for search engine algorithm:

1. Crawling: The process of gathering data

2. Indexing: Process of storing and tabulating the gathered data

3. Ranking: Selecting the best data to showcase to a user when searched

Getting the right keywords for your company and brand is a major task, brainstorming always helps to envision what might be searched or asked to get data by the user, stepping into the patient or caregivers' perspective also helps in case of OTC products, for prescription products a physician's perspective is very important. This data can be gathered from the KOLs and Influencers.

9. **Social Media Optimization & Marketing:** A key phase in the pharmaceutical industry is the use of social media marketing platforms like Facebook, Twitter, Instagram, Twitter, Linkedin, and Youtube. These platforms are growing rapidly as a key point for information capture and data points for the consumers. This also gives pharmaceutical marketers a strong opportunity to directly

interact with existing and potential customers by sharing product information, availability, cause, and effect, reach, and dialogue.

The social landscape is ever-changing, with the addition of the Metaverse, this space has become more exiting and expanded its horizons to offer a virtual reality world. Metaverse is the concept of an interconnected digital experience that complements and works in tandem with the physical world to create a fully immersive experience. The pharmaceutical companies can leverage Metaverse for healthcare education, engaging with patient communities and enhancing brand interaction.

Paid social media is the simplest form of marketing with the creation of community pages, product and corporate pages that are followed by users and contributions are by key stakeholders. This collaborative effort with KOLs and Influencers yields a lot of goodwill and trustable content.

10. **Audio Visual (AV) Marketing:** This is one of the most underused marketing platforms in the pharmaceutical Industry, it provides wonder and engaging way to showcase pharmaceutical product launches with the stakeholders, conduct CMEs and events with physicians, and make special announcements on the product and corporate front.

In nutraceuticals and other healthcare-related domains, the AV platform can be extensively used to generate user testimonials and get product betterment insights.

Paid visual advertising today compares to any mass media or television advertising, with great analytics and insight into the consumption behavior of such content.

Youtube and Vimeo are typically the most commonly used AV platform, with most of the social media sites allowing users to add backlinks to these channels or directly upload video content to these platforms.

11. **Direct Messaging – the Whatsapp effect:** Direct messaging has been a tool that has long existed as an alternative to SMS service. This has gained immense popularity in India in recent years with the penetration of mobile devices.

It has also made sharing of information and content easy and replies are almost instant and in real-time. With the introduction of Whatsapp Business, the direct messaging forum opened doors for showing business to a wider range of customers and audiences.

Whatsapp has been majorly used in the pharmaceutical industry for CRM purposes and interactions, information sharing with field forces with the creation of groups is also and very strong component for marketers. Other messaging platforms are commonly used in India like Telegram, Messenger, and Signal which have similar capabilities, but a less popular.

The main advantage of Whatsapp is to give you access to the phone number of the other person, and hence the advantages to not only customize a communication thread but also personalize it to a large extent. Integration of Whatsapp to the Facebook and Instagram pages or even to the corporate or product website is the next big thing to keep the consumer engaged and have access to the company at any given point in time.

12. **Content Marketing & Webinars:** Content is an embedded part of marketing, be it on the visual aid or in a communication material like leave behind, it has been the key conversation starter. In the post-pandemic world, Webinars have started playing their part in the sharing of knowledge, training, development, and beyond-borders conversations. The pharmaceutical industry has been part of this sea change in the creation and sharing of information, via different channels. Publishing high-quality product and indication content has helped many companies establish thought leadership in their key therapeutic areas.

   Perfectly curated content helps in establishing strong SEOs and has the potential to boost rankings. This is the best organic effort the pharmaceutical industry can deliver.

13. **Paid Search Results:** The Search engine and Social media visibility increase drastically with the infusion of budgets in the equation, traditionally we have trends that say pharmaceutical marketing budgets can range from as low as 5% to as high as 25%, most of these budgets go towards sales and promotion, a very small part of the budgets have been allocated to digital, over the years. This trend changed in 2020 when pharmaceutical brands allocated almost 66% of their budgets toward digital marketing. According to the report, COVID has driven digital marketing capabilities that were previously deprioritized to the forefront. It also has resulted in the realignment of organizations and processes to be more responsive to the customer, business, and environmental needs. To be seen and discovered, the pharmaceutical industry today has been actively bidding for certain keywords and purchasing advertising space in the search engine results.

The 2 main types of paid search ads are:

1. PPC, pay per click

2. CPM cost per mille

With PPC, companies only pay when they either run a campaign or if someone clicks on the link/ advertisement while CPM is based on the number of impressions. Google Adwords is the most widely used paid search advertising platform; however, other search engines like Bing, Yahoo, Ask, Duckduckgo, Alexa, and Linkedin also have paid programs. *(https://www.businesswire.com/ news/home/20210329005069/en/On-Average-Pharma-Brands-Allocated-66-Percent-of-Their-Marketing-Budgets-to-Digital-in-2020)*

14. **Blogs and Forums – the Quora Effect:** Subject matter experts in the pharmaceutical industry contribute to the creation of specific content on key therapeutic segments that are well researched and published. Placing blogs and write-ups across forums optimizes search results for the brand and builds trust and recognition. This web asset is a long-term strategy where stakeholders, consumers, and industry leaders can connect and interact. While pharmaceutical blogs and articles are published, forums initiate discussions and interactions, and the voice of the consumers can be heard from these channels. Quora clearly has been a popular multimedia forum where users pose questions and queries, getting answers from other individuals and even brands. This opens the most important channel of engagement and helps understand user behavior.

15. **Marketing Automation:** Multitasking for pharmaceutical marketers is the key, there is a limited amount of time to carry out tasks and give a return on investment matrix to the internal stakeholders. Automating advertising tasks is a programmatic way of aligning digital advertising as a function.

    This marketing automation function is also used on social media platforms to post relevant content at a predetermined schedule in sync with the organization's calendar. This avoids the need to be always present and manually share content and provides greater control over campaigns and the timing. Marketing and programmatic automation is the key to optimal resource utilization in pharma marketing, with consistent results and targets.

16. **PR and Reputation marketing:** Public relations (PR) and reputation marketing focus on gathering and promoting positive online reviews and published articles via trusted websites and news

sources. Companies with strongly positive reviews always has the potential to garner better employees, feature in the consumers' purchase decisions, and organically grow and build brands. Reading online reviews can influence customer buying decisions and is an important component of your overall brand and product strategy. Nutraceutical and Pharmaceutical companies with D2C brands rely heavily on an online reputation marketing strategy, which encourages customers to leave positive reviews on e-pharmacy & e-commerce sites where potential customers search for reviews.

THE APP ECOSYSTEM

App-based diagnostics

Smartphone-based clinical diagnostics is advancing through multiple IoT (internet of things) devices, today smart watches can measure heart rate and SPO2 (oxygen saturation) levels with ease, some smart devices also come with basic ECG capabilities that can help in pre-diagnosis of certain conditions. Diagnostic apps today help patients keep track of lifestyle diseases, maintain health history and records, and take consultation services with diagnostic tests and reports.

Some examples of applications are given below:

1. Phable - is an all-in-one wellness app that simplifies health management with intelligent insights and quick solutions for you. Phable enables its users to manage diabetes, hypertension, and other chronic conditions with a team of experts.

2. Eka Care - is a Connected Health care platform that is working towards developing a better healthcare system and a one-stop solution for all your family's health needs by connecting to the Medical health infrastructure, seamlessly

3. Medibuddy, Practo, Mfine, and Olivia are consolidators of physicians, healthcare consultants, and laboratories. They also provide an array of value-added services

VIDEO MEDICINE AND CONSULTATION (TELEMEDICINE)

Telemedicine as a term was coined in the early 1970s by Thomas Bird, which translates to "Healing at a Distance". From a simple phone-call-based consultation with a physician, the remote consultation scenario has gained immense popularity in times of highly communicable diseases and the current pandemic.

The Indian Telemedicine Market stood at US\$ 1314.83 million in FY2021 and is expected to grow at a robust CAGR of around 22.31%. This can be ascribed to the improving healthcare IT infrastructure across the country coupled with increasing digitization in India. Additionally, the growing prevalence of chronic and infectious diseases in the country with a shortage of healthcare professionals and physicians is further expected to support the market growth.

With a healthy competitive landscape companies like Practo, Apollo, Lyberate, and Doconline are making great reach and progress to the ever-increasing health-conscious population coupled with supportive government and insurance policies *(https://www.prnewswire.com/news-releases/india-telemedicine-market-report-2021-301396123.html)*.

ENGAGEMENT STRATEGIES FOR HEALTHCARE PROFESSIONALS

Digital engagement technologies open up a whole new world for marketing, the exchange of information, and even recruitment for clinical trials. Pharmaceutical sales reps, medical-science professionals, and patient-service teams can inform and influence physicians, and healthcare professionals in person - via mobile devices, video conferences, apps or social media. Patients are already starting to use patient portals for their medical records and to communicate with their physicians, and they use apps to hold medical data and information and online patient communities to speak to other patients with the same disease. Anytime-anywhere style of access to virtual care will become increasingly commonplace.

PATIENT ENGAGEMENT PROGRAMS

Today's patients are increasingly able and willing to take greater control of their own health, with the empowerment by the vast amount of health information available online and on apps, and by the array of health and fitness wearable. In a McKinsey survey, more than 85 percent of patients said they were confident in their ability to take responsibility for their health and knew how to access online resources to help them do so. Patients are becoming keener to evaluate different healthcare products and services, like preventive care, IoT-based diagnosis, and many more initiatives. In a digital world, the ability to engage with patients as they make such evaluations could be the key to the success of a pharmaceutical company's commercial model *(https://www.mckinsey.com/industries/life-sciences/our-insights/the-road-to-digital-success-in-pharma)*.

DOCTOR-PATIENT NETWORK/FORUMS

Prior to the digital era, pharma companies used to control the generation and dissemination of information about their products. Digital technologies have weakened that control, opening an array of new, independent information channels, forums, and networks.

Physicians today are a strong part of these interactions, with sources like WebMD or Mayoclinic gaining popularity for medical interactions, symptom lists, new treatment parameters and interactions.

Similar to physicians' interactions, there are online communities dedicated to sharing and discussing patients' experiences, apps and sensors to monitor the impact of therapy on a patient's life and wellbeing, and advanced data aggregation and analysis to generate new insights into drug safety and efficacy. The new generation of pharmaceutical companies will have to build the capabilities to anticipate or react rapidly to these new sources of evidence and also own and be responsible for the performance of their products.

MEASURING EFFECTIVENESS - DATA & ANALYTICS

UNDERSTANDING DIGITAL DATA

Big data is a large pool for data mining in the digital world; they form the raw material and basis for understanding the direction of decisions and strategy. The pharmaceutical industry usually sits on a wealth of raw data, which is hidden in different technical and organizational silos. Linking of this large pool of information to Data-driven insights can create a plethora of opportunities across domains.

From a pharmaceutical marketers' perspective, analytics of prescribers' behavior, geographies, and supply-demand parameters can help the sales teams understand the forecast and availability scenarios. Data from insurance claims, hospitals, labs and diagnostics can provide insights into the market of certain products in that geography. There are many automation tools available to support decision-making and the logic behind the decision.

The world is moving from "customization" to "personalization", which is suited for patient-specific dosing programs, where data will play a crucial role in the near future.

TOOLS AND TRICKS

Post-pandemic, pharmaceutical companies have started building digital capabilities, talent, and resources that are fragmented across smaller initiatives, rather than long-term BHAGs. Spending energies into 100s of smaller time-consuming strategies always yields mediocre results, it is always better to drive this through senior leadership with an outcomes-driven approach that can help the organization grow.

There is a range of free and paid tools available with varied functionality and customization options. The most widely used tool is Google Analytics, however other tools like Oribi, Monsterinsigts, Adobe Analytics, Mixpanel, Optimize, Hotjar, Matmo, and Kissmetrics are also quite famous. The use of the tool to automate your digital strategy and check the target and outcomes with checks and balances makes monitoring and evaluation a simpler process.

NEAR FUTURE DIGITAL OUTLOOK

Blockchain authentication against counterfeits

The pharmaceutical industry is growing at a rapid pace, with large investments and demand for products not only in India but across the globe. As the brands in pharmaceuticals grow by value and volume, it is evident that the counterfeiting industry makes its way to cull part of the profit and extract money from the legitimate supply chain. To counter this, technology companies have come up with multiple solutions from a simple QR bases authentication system to complex Blockchain methods. The block chain method is interesting as pharmaceutical manufacturers can now also authenticate APIs, Excipients, and other relevant ingredients to their source and date of manufacturing. This involves supply chain embedded parameters to track and trace product origins.

AI-based supply chain initiatives

With AI-based prediction models now embedded in the current generation ERP systems, pharmaceutical companies can now do the following:

- Visibility of products across the supply chain
- Counterfeit avoidance strategies

- Cost efficiency in supply chain operations
- Demand forecasting for optimizing production
- Supply chain software deployment support
- Packaging and track and trace strategies
- Track and trace managed services
- Sustainability and re-commerce focused track and trace strategies
- Circularity protocols

These initiatives make it easier for marketers to communicate to the physicians and stakeholders about the product authenticity and measures taken to keep it watertight *(https://tongadive.com/)*.

DEEP INTELLIGENCE AND MACHINE LEARNING

Understanding the buying behavior of consumers to seasonal consumption of drugs according to ailments, all are part of the planning and initiative for a marketer. The technological advancement in gathering this data from historical models, with predictions from the healthcare ecosystem - all combined given us deeper insights about products and services that might be in demand and have a spike in consumption. This is an ever-growing technology and evolving. Industry 4.0 will be based on ML and Deep Intelligence for pharmaceutical manufacturing

SUMMARY

This chapter deals with digital ecosystem, components of digital marketing, digital strategies for engaging with healthcare professionals & patients, measuring the effectiveness of digital marketing and the near future digital outlook. Key highlights are as follows-

1. Digital ecosystems comprise multiple stakeholders and are based on three major pillars - i) Digital Infrastructure, ii) Digital Data, Social and Security & iii) Digital Transactions and the Economy

2. Components of digital marketing include Websites, Online Advertising, Email Marketing, Mobile Marketing, In-App Marketing, Influencer Marketing, Retargeting & Remarketing, Search Engine Optimization & Marketing, Social Media Optimization & Marketing, Audio Visual (AV) Marketing, Direct Messaging – the Whatsapp effect, Content Marketing & Webinars, Paid Search Results, Blogs and Forums – the Quora Effect, Marketing Automation & PR and Reputation marketing

3. Smartphone-based clinical diagnostics is advancing through multiple IoT (internet of things) devices. These apps are being used for patient engagement and engagement with the doctors.

4. Pharmaceutical sales reps, medical-science professionals, and patient-service teams can inform and influence physicians, and healthcare professionals in person - via mobile devices, video conferences, apps or social media.

5. Today's patients are increasingly able and willing to take greater control of their own health, with the empowerment by the vast amount of health information available online and on apps, and by the array of health and fitness wearable. In a digital world, the ability to engage with patients as they make such evaluations could be the key to the success of a pharmaceutical company's commercial model.

6. From a pharmaceutical marketers' perspective, analytics of prescribers' behavior, geographies, and supply-demand parameters can help the sales teams understand the forecast and availability scenarios. Data from insurance claims, hospitals, labs and diagnostics can provide insights into the market of certain products in that geography. There are many automation tools available to support decision-making and the logic behind the decision. The world is moving from "customization" to "personalization", which is suited for patient-specific dosing programs, where data will play a crucial role in the near future.

7. There is a range of free and paid tools available with varied functionality and customization options. The most widely used tool is Google Analytics, however other tools like Oribi, Monsterinsigts, Adobe Analytics, Mixpanel, Optimize, Hotjar, Matmo, and Kissmetrics are also quite famous. The use of the tool to automate your digital strategy and check the target and outcomes with checks and balances makes monitoring and evaluation a simpler process.

8. Blockchain authentication against counterfeits and AI-based supply chain initiatives will make it easier for marketers to communicate to the physicians and stakeholders about the product authenticity and measures taken to keep it watertight.

REFERENCES

1. Uplers Email - https://email.uplers.com/blog/email-marketing-for-the-pharma-industry/

2. https://www.aha.io/roadmapping/guide/marketing-plans/digital-marketing-plan-components

3. https://www.businesswire.com/news/home/20210329005069/en/On-Average-Pharma-Brands-Allocated-66-Percent-of-Their-Marketing-Budgets-to-Digital-in-2020

4. https://www.prnewswire.com/news-releases/india-telemedicine-market-report-2021-301396123.html

5. https://www.mckinsey.com/industries/life-sciences/our-insights/the-road-to-digital-success-in-pharma

6. https://tongadive.com/

CHAPTER **19**

HOSPITAL (INSTITUTIONAL) MARKETING

LEARNING OUTCOMES

After reading this chapter, you should be able to:

- Understand the significance of hospital/institutional marketing

- Understand the institutional marketing process

- Understand institutional marketing strategies

SIGNIFICANCE OF HOSPITAL/INSTITUTIONAL MARKETING

When pharmaceutical companies or their authorized agents supply or contract to supply pharma products directly to any hospital, medical college, corporate organizations, or central & state government institution is known as the institutional sale. Institutional sales could also be termed Business to Business (B2B) in pharmaceutical marketing.

The Indian healthcare sector is one of the fastest-growing sectors and is witnessing an expansion by existing hospitals in terms of new services through technology platforms, investment in next-generation tools for diagnosis, tie-ups with insurance companies, and geographical expansion, with a focus to provide world-class medical care at affordable cost. An estimated 69 thousand public and private hospitals were recorded across India in 2019. Of these, 43 thousand were private sector hospitals (Sanyukta Kanwal, Mar 17, 2022, www.statista.com).

The challenges are ever-growing and in markets that have become dynamic, pharmaceutical companies are continuously reinventing themselves for better business opportunities. The institutional or direct business is one such aspect, where large-format institutions like corporate & network hospitals, medical colleges, government hospitals, co-operative pharmacy chain's governing bodies, online pharmacies fulfillment hubs, etc., purchase medicines in the bulk quantity at a predefined rate contract for a fixed term. This is different from a traditional form of pharmaceutical distribution and is a more direct line of business-to-business (B2B) products. It is estimated that the hospital market in India is approximately 9 percent of the total pharmaceutical market.

THE INSTITUTIONAL MARKETING PROCESS

Earlier, pharmaceutical companies used to direct their sales team leaders to sell to institutional players and fill tenders via distributors. But in recent years most large and mid-size pharma companies have started maintaining a separate institutional sales force to handle this business as a separate profit center. Institutional sales teams are responsible for handling and covering all major government, corporate, and chain pharmacies, marketing, direct handling of some aggregators, responsibilities for tender sales, and relationship management with tender committees.

In pharmaceutical institutional sales, individual physicians or persons are not decision-makers but decision making is by multiple stakeholders, preceded by the management committee. This makes it different from regular pharmaceutical marketing to doctors who decide what to prescribe. In some cases, the decision about which drugs are to be purchased is made by a committee comprising of physicians, pharmacists, nurse practitioners, and others. Most of the purchase decisions are based on product portfolio, scientific research, clinical evidence, expert opinions of KOLs, and cost-benefit studies with a deep understanding of the manufacturing company and quality parameters.

INSTITUTIONAL MARKETING STRATEGIES

Institutional business for pharma is directly proportional to the right product and promotion mix by the field force. Until there is critical mass in the product and visibility with KOLs, most hospitals do not recommend a brand, with a strong presence and active prescriptions, popularity also matters, in this case, to make to the recommendations committee.

Once the committee suggests the brand and molecules, the integration in the tender document or procurement process takes place. This is followed by contacting the manufacturer or brand owner to supply the said product directly to the institution at the best possible price point for a fixed quantity and with a strict timeline.

The distribution channel for this B-to-B orientation is owned and operated by the institution, company, or aggregator, the main thought is backward integration, with no or low sharing of margins among other members of the distribution channel. These companies procure medicines and other pharmaceutical products through offering tenders, tie-up directly with pharmaceutical manufacturers, large format distributors, or C&Fs. This helps them to minimize the cost of goods procured and maximize profitability, also pass on the benefit to consumers and patients.

In some instances, the pharma companies are also directed to mention the specific channel the product has been manufactured to be distributed from, for example in the case of the Pradhan Mantri Jan Aushadhi Kendra, some packs mention the product is specifically manufactured for sale only through the PMJAK channel.

The government e-marketplace (GeM) was launched in Aug 2016 to reduce manual process inefficiencies and human interventions in public procurement. GeM has 20,000 plus government organizations registered as buyers (Government E-Marketplace (GeM) Handbook, July 2018).

FUTURE SCOPE

In the ever-evolving pharmaceutical industry, the number of corporate, multi-specialty hospitals and centralized institutions is becoming more common, and the smaller hospitals and chain clinics could see an acquisition spree or brand consolidation soon. This brand-driven healthcare ecosystem is creating a huge scope in institutional sales for the industry.

SUMMARY

This chapter deals with the significance of hospital/institutional marketing, the process of institutional marketing, and institutional marketing strategies. Key highlights are as follows: -

1. When pharmaceutical companies or their authorized agents supply or contract to supply pharma products directly to any hospital, medical college, corporate organizations, or central & state government institution is known as the institutional sale. Institutional sales could

also be termed Business to Business (B2B) in pharmaceutical marketing.

2. An estimated 69 thousand public and private hospitals were recorded across India in 2019. Of these, 43 thousand were private sector hospitals. It is estimated that the hospital market in India is approximately 9 percent of the total pharmaceutical market.

3. In pharmaceutical institutional sales, individual physicians or persons are not decision-makers but decision making is by multiple stakeholders, preceded by the management committee. Most of the purchase decisions are based on product portfolio, scientific research, clinical evidence, expert opinions of KOLs, and cost-benefit studies with a deep understanding of the manufacturing company and quality parameters.

4. Institutional business for pharma is directly proportional to the right product and promotion mix by the field force. The distribution channel for this B-to-B orientation is owned and operated by the institution, company, or aggregator.

REFERENCES

1. Kanwal Sanyukta, Estimated number of public and private hospitals in India 2019, Mar 17, 2022, www.statista.com https://www.statista.com/statistics/1128425/india-number-of-public-and-private-hospitals-estimated/#:~:text=An%20estimated%2069%20thousand%20public,number%20of%20hospitals%20that%20year accessed on April 20, 2022

2. Government E-Marketplace (GeM) Handbook, July 2018

CHAPTER 20

RURAL MARKETING – ITS SIGNIFICANCE IN THE INDIAN CONTEXT

LEARNING OUTCOMES

After reading this chapter, you should be able to:

- Understand the rural healthcare scenario in India
- Understand the significance of rural marketing in the pharmaceutical industry
- Learn strategies for rural marketing

INDIAN RURAL HEALTHCARE SCENARIO

Around 65.5 percent (89.5 crores) of the Indian population lives in rural areas (estimated mid-year 2021 population, Rural Health Statistics 2020-21, Government of India, Ministry of Health and Family Welfare) in 6,62,050 villages.

The availability of healthcare in rural India is still very underdeveloped. To increase access to healthcare in rural areas, the health care infrastructure has been developed as a three-tier system based on the population norms as follows (Rural Health Statistics 2020-21, Government of India, Ministry of Health and Family Welfare):

- Sub Centres (SCs)

- ✓ The Sub Centre is the most peripheral and first contact point between the primary health care system and the community. Sub Centres are assigned tasks relating to interpersonal communication to bring about behavioral change and provide services in different programmes like maternal and child health, family welfare, nutrition, immunization, diarrhoea control, and communicable diseases as well as non-communicable diseases.
- ✓ A total of 156101 Sub Centres are functioning in rural areas of the country as of 31st March 2021

- Primary Health Centres (PHCs)
  - ✓ PHC is the first contact point between the village community and the medical officer. The PHCs were envisaged to provide an integrated curative and preventive health care to the rural population with emphasis on preventive and promotive aspects of health care.
  - ✓ 25140 PHCs are functioning in rural areas of the country as of 31st March 2021

- Community Health Centres (CHCs)
  - ✓ A CHC is required to be manned by four medical specialists i.e. Surgeon, Physician, Obstetrician/Gynecologist, and Pediatrician supported by paramedical and other staff (See Annexure-I for IPHS norms). It has 30 in-door beds with one OT, X-ray, labour room, and laboratory facilities. It serves as a referral centre for 4 PHCs and also provides facilities for obstetric care and specialist consultations.
  - ✓ As of 31st March 2021, there are 5481 of CHCs functioning in rural areas of the country.

GOVERNMENT INITIATIVES

- The National Rural Health Mission (NRHM) was launched to provide accessible, affordable, and quality health care to the rural population, especially the vulnerable groups.

- Ayushman Bharat to provide a coverage of up to 5 lakh INR to 10 crore poor families is the biggest scheme of its kind in the world and is in continuation with the trend of the government being a payer rather than a provider in the secondary and tertiary care space. The National Health Protection Scheme will provide much-needed protection to the most vulnerable section of our population

The Ayushman Bharat - Health and Wellness Centres (AB-HWCs) were launched under the Ayushman Bharat Programme in a bid to move away from selective health care to a more comprehensive range of services spanning preventive, promotive, curative, rehabilitative, and palliative care for all ages. The National Health Policy of 2017 envisioned these centres as the foundation of India's health system. The service delivery including preventive, promotive, curative, and rehabilitative health care is at three levels i.e. i. Family/Household and community levels through outreach OPDs, Health Mela, Village Panchayat, Village and Home Visits, School and Anganwadi visits ii. AB-HWCs iii. Referral Facilities/Sites: Delivery of services closer to the community and close monitoring would enable increased coverage and help in addressing issues of marginalization. As of 14th Sept 2021, there are 77,784 AB-HWCs. (Ayushman Bharat Health and Wellness Centres booklet, file:///C:/Users/Vijay/Downloads/Reforms_Booklet_HWC_English_updated_14th_Sep_2021.pdf)

- **e-initiatives:** The Ministry of Health and Family Welfare (MoHFW) is also running several e-governance initiatives for the digitization of the healthcare sector in India and has set up a division called e-Health India. e-Health adds two more goals to the triple aim of healthcare globally:

  - Access
  - Affordability
  - Quality
  - Lowering disease burden
  - Efficient monitoring of health entitlements to citizens

SIGNIFICANCE OF RURAL MARKETING IN THE PHARMACEUTICAL INDUSTRY

Opportunities

Nearly 75 percent of dispensaries, 60 percent of hospitals, and 80 percent of doctors are located in urban areas (KPMG OPPI report on healthcare access initiatives, Aug 2016). Thus, there is tremendous scope for the growth of healthcare in rural areas. The Indian government aims at increasing healthcare spending to 2.5 percent of the GDP (gross domestic product) by 2025. Remote healthcare, patient empowerment, and multichannel engagement are likely to be some of the important emerging trends in the sector due to the pandemic. Increasing penetration of health insurance and pharmacies is expected to drive expenditure on medicine.

With 65.5 percent of India's population residing in rural areas, pharmaceutical companies have immense opportunities to tap into the rural market. Demand for generic medicines in rural markets has seen a sharp growth and various companies have already invested in expanding the distribution network in rural areas.

With the growth in incomes, increasing health awareness and greater penetration, the rural contribution to total pharma revenue is expected to grow from 20% to approximately 25% by 2020, according to a 2011 report by India Infoline. It could prove to be a key revenue driver in the near future (Meenakshee Sinha and S. Arunachalam, Forbes, April 16, 2019).

Challenges

- Poor healthcare infrastructure - very low doctor to patient ratio
- Healthcare delivery – modern medicine is not available, most primary care physicians are Ayurvedic or Unani practitioners
- Communication barrier due to poor literacy rates
- Low healthcare awareness
- Price sensitivity
- Making products available
- The paucity of talent for marketing in rural areas

RURAL MARKETING STRATEGIES IN THE PHARMACEUTICAL INDUSTRY

Many Indian companies like Mankind, Alkem, Lupin, and Sun as well as multinational companies like Sanofi, Novartis, Abbott, GSK, Merck, and Eli Lilly are focusing on rural markets for their growth. Some of the key rural initiatives are stated below: -

1. **Novartis Arogya Parivar:** Novartis recruits and trains locals in remote villages to become "health educators," who help inform communities about health, disease prevention, and the benefits of seeking timely treatment. Local teams work with doctors to organize health camps in remote villages – mobile clinics that provide access to screening, diagnosis, and therapies. From 2010 onwards, outreach in rural areas across 11 Indian states has brought health education to more than 36 million people and direct health

benefits to 2.5 million patients through diagnosis and treatment. Arogya Parivar is a leading example of shared value (www.Novartis.in).

2. **Sanofi Prayas:** Sanofi-aventis India launched Prayas aimed at bridging the diagnosis-treatment gap through a structured continuing education program for rural doctors across India. Through Prayas, specialists from semi-urban areas share the latest medical knowledge and clinical experience with general practitioners based in smaller towns and villages in the interiors of India. Sanofi-aventis' second strategy for improving healthcare access is to make quality medicines available at affordable prices for rural Indian patients. Sanofi-aventis aims to extend the benefits of Prayas to as many as 100,000 doctors in rural areas and thus contribute to India's much-needed healthcare capacity building initiative (www.Sanofi.in).

3. **Mankind Pharma:** Mankind pharmaceutical was launched in 1995 and focused on rural areas, areas that were not covered well by other pharmaceutical companies. They targeted lower and price-sensitive segments in rural areas. They launched quality medicines at a ruthlessly competitive price that other big pharmaceutical companies found difficult to match. The strategy was to go after the masses with products they can afford. They focused on flat field structure. They gave heavy incentives to their field staff and distribution partners. The aggressive sales force meticulously reached places where others couldn't, making it one of the fastest-growing pharma companies within a decade (The Economic Times, Jun 13, 2012).

The rural marketing strategies followed by pharmaceutical companies are as follows: -

- Build trust among the population
  - ✓ Providing quick relief is the key
- Customized offerings
  - ✓ Appropriate pack size and a price point
- Focus on 4A's –
  - ✓ Accessibility
    - Identify doctors capable of diagnosing and treating
    - Physical distribution network to ensure the availability of medicines, and even tie-ups with e-pharmacies can help

- ✓ Affordability
  - ▪ A lower price point is important as there is price-sensitivity
- ✓ Awareness
  - ▪ Create disease awareness
  - ▪ Problem-solution approach
  - ▪ Use local communication channels
  - ▪ Use local language
- ✓ Acceptability
  - ▪ Product formulated in a manner that the rural audience sees value and it is readily accepted
- • Dedicated team to market in rural areas
  - ✓ Identify local talent

The Healthcare sector, be 'Pharmaceuticals' or 'Healthcare service providers', has tremendous potential to grow in rural India. Many pharmaceutical companies & healthcare service providers are now focusing on rural areas. They are investing in products specially designed for rural markets, carrying out promotional campaigns in rural areas (be it education or awareness drive), and increasing access to products in rural areas & there is a huge opportunity, healthcare marketing in rural India also poses lots of challenges. Those players who have taken initiative & are continuously harping on rural markets with innovative tailor-made marketing activities are going to be the winners in the long run (Vijay Bhangale, Jan 2012)

SUMMARY

This chapter deals with the rural healthcare scenario in India, the significance of rural marketing, and rural marketing strategies in the pharmaceutical industry. Key highlights are as follows: -

1. Around 65.5 percent (89.5 crores) of the Indian population lives in rural areas in 6,62,050 villages. The availability of healthcare in rural India is still very underdeveloped.

2. Nearly 75 percent of dispensaries, 60 percent of hospitals, and 80 percent of doctors are located in urban areas. Thus, there is tremendous scope for the growth of healthcare in rural areas.

3. With 65.5 percent of India's population residing in rural areas, pharmaceutical companies have immense opportunities to tap into the rural market. Demand for generic medicines in rural markets has

seen a sharp growth and various companies have already invested in expanding the distribution network in rural areas.

4. With the growth in incomes, increasing health awareness and greater penetration, the rural contribution to total pharma revenue is expected to grow from 20% to approximately 25% by 2020, according to a 2011 report by India Infoline. It could prove to be a key revenue driver in the near future.

5. Challenges for rural marketing in pharma are poor healthcare infrastructure - very low doctor to patient ratio, modern medicine not available - most primary care physicians are ayurvedic or unani practitioners, communication barrier due to poor literacy rates, low healthcare awareness, price-sensitivity, making products available and paucity of talent for marketing in rural areas.

6. Rural marketing strategies adopted by pharmaceutical companies are building trust among the population, providing customized offerings, focusing on 4A's – Accessibility, Affordability, Awareness and Acceptability & having a dedicated team to market in rural areas.

REFERENCES

1. Rural Health Statistics 2020-21, Government of India, Ministry of Health and Family Welfare available at https://main.mohfw.gov.in /sites/default/files/rhs20-21_2.pdf accessed on April 20,2022.

2. Ayushman Bharat Health and Wellness Centres booklet, file:///C:/Users/Vijay/Downloads/Reforms_Booklet_HWC_English_u pdated_14th_Sep_2021.pdf.

3. Sinha Meenakshee, Arunachalam S., Forbes, April 16, 2019.

4. https://www.novartis.in/our-work/expanding-access-healthcare/novartis-social-ventures

5. https://www.sanofi.in/dam/jcr:e798fb8d-97bb-49e9-b719-111735c6eda9/Sanofiaventis%20launches%20PRAYAS%20an%20 endeavor%20to%20meet%20rural%20Indias%20healthcare%20ne eds.pdf.

6. Rajagopal Divya, How Ramesh Juneja's Mankind Pharma has changed pharma game with pulp marketing, The Economic Times, Jun 13, 2012. Bhangale Vijay, Healthcare marketing Initiatives in Rural India, Vision Research, Vol. 2, No.1. Jan 2012.

INTERNATIONAL MARKETING – EXPLOITING THE POTENTIAL OF GENERICS

LEARNING OUTCOMES

After reading this chapter, you should be able to:

- Understand International marketing and the international marketing environment
- Gain knowledge about the global pharmaceutical market and the role of the Indian pharmaceutical industry
- Global pharmaceutical market classification
- Selection of international market
- Get an insight into the internationalization of the Indian pharmaceutical industry
- Learn how the marketing mix for international marketing is different
- Understand how to register products in international markets

INTERNATIONAL MARKETING

International marketing is defined as the marketing activities carried out across national boundaries. The basic principles of marketing remain the same whether it is domestic or international marketing. However, differences in factors like political, economic, socio-cultural, legal, technological & environment make international marketing a distinct discipline (Rakesh Mohan Joshi, 2014). While conducting international marketing, firms will have to deal with activities across various countries, foreign exchange, duties & tariffs, different cultural environment, customer preferences, human resources, and the risks associated with these changes. In the case of the pharmaceutical industry, there is an additional tough requirement to deal with i.e., regulatory affairs & control systems of each country.

INTERNATIONAL MARKETING ENVIRONMENT

In international markets, the marketer faces environmental factors which are beyond the organization's control and it makes the task challenging when compared to domestic marketing. If an organization is operating in several markets, then the marketing complexities increase. These environment factors (adapted from the book International Marketing, Rakesh Mohan Joshi, 2014) are discussed below:

Political: The country's political stability & its government's policies greatly influence the international markets. A marketer would look for a country to be politically stable & having favorable business policies. E.g., At present, the countries like Afghanistan, Syria & Myanmar have a lot of uncertainty. The ongoing war between Russia & Ukraine and economic sanctions on Russia, it is affecting the region. There is virtually no trade between India & Pakistan due to political differences.

Economic: The economic stability in the target markets is equally critical for success in international markets. Economic uncertainties and hyperinflation can create severe problems regarding the certainty of payments. The soundness of financial institutions like banks is equally important. Since marketers deal in foreign exchange, the forex reserves of a country & healthy status of the country's balance of payment are most important. E.g., Due to hyperinflation in countries like Venezuela, Argentina & Zimbabwe, their economies were in the doldrums. At present Sri Lanka & Pakistan are facing problems of depleting foreign exchange. This has greatly affected the country's ability to make payments. A few years back the economy of Greece collapsed due to the huge burden of foreign loans. Such countries provide a great risk to marketers & should be avoided.

Socio-Cultural: Cultural factors play an important role when targeting international markets. The culture of target markets affects product modifications, especially in the case of consumer products. The social environment also affects the motives to make buying decisions & communication strategies need to be customized for different markets. The social beliefs also vary among countries and the marketing mix has to be tailor-made to suit the social norms of the target market. The cultural factor includes language, religion, customs & traditions, lifestyles, consumption habits, family norms, gestures &business etiquette. E.g., on religious grounds, insulin made from porcine (pork) source can't be supplied in Middle East countries, while that from a bovine (cow) source can't be supplied in India. Vegan foods are popular in advanced countries. Hard gelatin capsule manufacturers are now producing 'vegetarian' capsules made from cellulose.

Legal: Every country will have its own set of laws & marketers will have to comply with the laws of the land. Some of the laws impacting international marketing are the Export-Import policy, customs law, GST/VAT, foreign exchange regulations, foreign investment laws, contract Act, income tax, and labour laws. From an Indian pharmaceutical company's point of view, important laws are the Food & Drugs law, Drug registration regulations, and promotion of drugs, production, imports, distribution, pricing & sale of pharmaceuticals.

Technological: There are vast variations in the availability of technology between developed and emerging markets. In developed countries, there is a high level of automation as compared to emerging markets where a lot of operations are done semi-automatic or manual. In many African markets, there is inadequate coverage of electricity, telecommunication & internet. This poses challenges for e-commerce & online payments. As per the government of India regulations, it is mandatory to have Bar Code on pharmaceutical product labels incorporating essential details like Batch No., Manufacturing date, Expiry date, Supplier's code. For developed countries goods are always packed in 'Pallets' & containerized.

Environmental: There is greater awareness of environmental issues across the world. Most countries have environmental protection laws & regulations. These laws cover issues related to air, water & land pollution. Protection of flora & fauna, industry pollution, management of e-waste, disposal of waste, reducing carbon footprint, environment-friendly products, conservation of resources, etc. The implications for the pharmaceutical industry are to comply with local laws, reduce the use of plastics in packaging, incorporate recyclable materials, manage effluent treatments, and manage industrial waste including biological.

GLOBAL PHARMACEUTICAL INDUSTRY

According to IQVIA Market Prognosis, Sep 2021; IQVIA Institute, Nov 2021, the global spending on medicines in 2021 is 1423.5 USD and is estimated to be nearly USD 1.8 trillion in 2026. The break-up of spending among different countries is as given below in Table 21.1.

Table 21.1 Break-up of global spending on medicines in 2021

Group	Spends in USD Billion
Developed countries	1050.4
Pharmerging (India is part of this)	354.2
Others	19.0
Global market	1423.6

The other key highlights of the global pharmaceutical market are as follows: -

- Among the world pharmaceuticals markets, North America was the largest, covering 46% of the global market in 2020. Asia Pacific region was the 2nd largest region covering 26%. Africa was the smallest region in the global pharmaceuticals market (Pharmaceuticals Global Market Report 2021: COVID-19 Impact and Recovery to 2030, Research and Markets, Dublin, March 31, 2021)

- Spending across major pharmerging markets is expected to grow 5-8% CAGR through 2026, with China slowing to 2.5-5.5% while Brazil, India, and Russia are expected to grow by more than 7.5% CAGR through 2026 (IQVIA report on The Global Use of Medicines 2022)

- Currently, India is ranked as the 11th largest market and is expected to be ranked as the 9th largest by 2026 (IQVIA report on The Global Use of Medicines 2022)

- The two leading global therapy areas — Oncology and immunology — are forecast to grow 9–12% and 6–9% CAGR through 2026, lifted by significant increases in new treatments and medicine use and offset by losses of exclusivity, including biosimilars (IQVIA report on The Global Use of Medicines 2022).

INDIAN PHARMACEUTICALS INDUSTRY & GENERIC MARKET

India's Pharmaceutical Industry in 2020-21 reached US$ 49 billion including Exports & Domestic (Annual Report 2020-21, Pharmaceuticals Export Promotion of India). The domestic market for pharmaceuticals is bigger than the market for exports. Nevertheless, due to the growing generics global market, strict price controls of the domestic market, as well as better margins, the market for exports is growing far better as compared to the domestic market. During the year 2020-21, India's pharma exports were US$24.47 bn. consisting of finished formulations, bulk drugs, surgical items, and Ayush & herbal products. (Annual Report 2020-21, Pharmaceuticals Export Promotion of India). Let's have a look at the generics market.

Table 21.2 Generics market region wise (FY-2021 estimates)

Region	USD Billion
Asia	156.4
Americas	89.5
Western Europe	54.51
Eastern Europe	25.88
Africa & Middle East	13.27
Total Market	339.56

Source: Annual Report 2020-21, Pharmaceuticals Export Promotion of India

During the year 2020-21 Indian Pharmaceutical industry manufactured generic forms of finished dosages worth US$40.85 bn. India manufactures generic drugs around 20-24 % of the total world's production by value (Annual Report 2020-21, Pharmaceuticals Export Promotion of India).

During FY-21 exports of generic drugs from India reached $18.85 bn. showing 19.53% growth which is almost six times the estimated growth rate of the global generic market which is the highest over the previous seven years. That makes India the biggest exporter of generics worldwide. During FY-21, there were 202 countries where India exported pharmaceuticals. Out of these, the USA market contributed over 31.6% of the total which is more than US$ 1bn. in value. Also, 52 countries have exported over US$ 100 million while another 15 countries have grown more than 30 percent (Annual Report 2020-21, Pharmaceuticals Export Promotion of India).

During FY-2021, 1438 market authorizations were granted by USFDA. Companies based in India received 36% of these authorizations. There are 741 pharmaceutical factories in India which are registered with USFDA. In 2019, eight Indian companies were featured in the list of top 20 generic manufacturers in the world based on revenues. India's vaccine manufacturers export to the tune of $140 million apart from meeting our government's commitments of millions of free doses supplied abroad. This is the testimony of our scientists 'capacity in meeting global demands with speed & high accuracy. (Annual Report 2020-21, Pharmaceuticals Export Promotion of India)

GENERIC FORMULATIONS – A COMPETITIVE ADVANTAGE OF INDIA

India has become a major global player in generics formulations. India has a competitive advantage as a nation in generic formulations due to the following favorable factors -

- State-of-the-art manufacturing facilities
- A large pool of technically qualified manpower
- Large raw material base & low cost of operations
- English language advantage

Following additional factors led to India's success:

1. Patent expirations of several branded drugs during the last few years.

2. Indian pharmaceutical companies making generics realised the importance of expiring patents of branded drugs and willingness to make substantial additional investments in drug development & research.

3. Due to pressure on healthcare providers to contain healthcare costs, there is a preference for generics.

4. High confidence in generics among consumers of developed countries.

5. Across the world, a large population is aging which required lower-cost treatments.

6. Outbreak of diseases like AIDS & Hepatitis in developing countries, requiring cost-effective medicines for large masses.

GLOBAL PHARMACEUTICAL MARKET CLASSIFICATION

The Pharmaceutical market based on the diversity in the regulation and marketing interests can be divided into two groups - Regulated markets and Emerging markets (Semi–regulated) (Badjatya et. al. 2013) comprising different countries as follows: -

1. **Regulated Market**: USA, EU (UK, Germany, France, Ireland, Sweden, etc.), Japan, Canada, Australia, New Zealand, South Africa

2. **Semi regulated Market**: (Rest of World or ROW Countries):

   (a) Asia (Sri Lanka, India, Bangladesh, ASEAN: 10 Countries group - Philippines, Vietnam Singapore, Malaysia, Thailand, Indonesia, Laos, Cambodia, Brunei Darussalam, Myanmar.

   (b) African countries (Algeria, Zambia, Ethiopia, Ghana, Kenya, Malawi, Mozambique, Namibia, Nigeria, Sierra Leone, Tanzania, Zimbabwe, etc.)

   (c) Middle East countries (Gulf Co-operation Council countries i.e., Bahrain, Kuwait, Oman, Qatar, Saudi Arabia, UAE)

   (d) Latin America (Mexico, Brazil, Panama, Peru, Guatemala, Argentina, Chile, Dominican Republic)

   (e) CIS (commonwealth of independent states): Russia, Ukraine, OFSUs (Armenia, Azerbaijan, Belarus, Georgia, Kazakhstan, Kirghizstan, Moldova, Tajikistan, Turkmenistan, Uzbekistan, etc.)

SELECTION OF INTERNATIONAL MARKETS

As firms have limited resources, they could focus only on a few overseas markets. Hence, proper market selection minimizes the wastage of a firm's resources & time. The process of market selection involves the evaluation of different market segments and concentrating marketing efforts on a country that has greater potential. The criteria for evaluation (adapted from the book International Marketing, Rakesh Mohan Joshi, 2014) are as follows-

1. **Market Size:** Estimate the current market size and its future potential. The USA is the largest pharmaceutical market followed by Europe & the ROW.

2. **Accessibility to international markets**: Any market having good potential & profitable may not become attractive because of various marketing barriers or financial controls as follows-

   (a) **Tariff barriers** like custom duties & taxes

    (b) **Non-tariff barriers** like state trading, customs entry procedure/ pre-shipment inspection, product safety requirements, quotas, and

    (c) **Financial controls** like the crunch of foreign exchange, credit restriction, restriction on profit repatriation, prior import deposit, and multiple exchange rates.

3. **Profitability**: For generic formulations, regulated markets of the USA, Europe & Japan are highly profitable than ROW.

Distance model of selection of international markets (adapted from the book International Marketing, Rakesh Mohan Joshi, 2014). International market selection can be arrived at by a combination of the distances given below-

1. **Geographical distance**: Nations having geographical proximity are natural target markets because of shorter physical distances and lower complexities in logistics. For lower valued goods neighboring countries are natural target markets. E.g., India has huge trade with Nepal, Srilanka & Bangladesh.

2. **Economic distance**: The landed cost of products in target markets, as well as ease of doing business, is decided by the economic distance. E.g., Trade with Afghanistan and Myanmar may not economical though the geographical distance is not far.

3. **Political distance**: The political relationship between two governments influences target market selection. Hostile political relations would stop trade with markets that may be attractive. E.g., India's trade with Pakistan.

4. **Psychic distance**: It is a psychological gap created due to communication barriers resulting from differences in awareness levels, lifestyles, the level of technical skills, cultural orientation, & political ideologies. As a result of the difficulty in obtaining information about foreign markets and uncertainty, the firm prefers to deal with countries having few barriers. E.g., India's trade with South American countries – Argentina, Venezuela.

INTERNATIONALIZATION OF
INDIAN PHARMACEUTICAL INDUSTRY

Globalization has generated a more competitive market scenario across the world. Therefore, firms should have competitive advantages to become globally successful. India like most emerging economies has gone through a fundamental change in its policy orientation from an import substitution role to an export-led outward economy. Indian pharmaceutical companies faced domestic competition from global firms and were required to upgrade their capabilities for survival.

In 1991, Govt. of India initiated liberalization and removed restrictions on imports, trade & FDI overseas which helped Indian companies in the process of international expansion and gave access to the Indian market to foreign companies. As a result of this liberalization, several Indian pharmaceutical companies started successfully their overseas operations and become major suppliers and producers of generic formulations worldwide. India became a signatory to the WTO order, there was a paradigm shift in the Indian pharmaceutical industry ushering in the Product Patent Regime. In 2005, the Indian patent act of 1970 was amended so that India became compliant with TRIPs (Trade-Related Aspects of Intellectual Property rights), an important for pharmaceutical products in India.

Many Indian pharmaceutical companies have further moved up in the value chain & thus internationalized by acquiring small firms and also setting up wholly-owned subsidiaries, to access resources and enter newer markets. Several Indian pharmaceutical companies are accessing developed markets and acquiring new technology following the process of internationalization. Indian pharmaceutical companies have duly adapted to the hard realities of globalization and are creating new niches via the internationalization process.

Three developments are pushing the expansion of the Indian pharmaceutical industry into overseas markets (adapted from Internationalisation strategies of Indian pharmaceutical firms, Dinar Kale, 2010) as follows-

A. **Hatch-Waxman Act & Opportunities for Generic Drugs in the USA:** Hutch Waxman Act was passed in the USA in 1984. As per this new law, generic drugs manufacturers would not have to pass through a prolonged period of extensive clinical trials for entering the USA market. Just the generic drug had to demonstrate bio-equivalence to original innovator Brands. It was enough to acquire a patent on a generic drug. Similarly, procedures were set up for dispute settlement between generic drug producers & branded drug producers.

   USA market provides a lucrative business opportunity for Indian pharmaceutical companies as they enjoy a low-cost advantage due to economies of scale, local manufacture of bulk drugs & relatively lower cost of skilled as well as scientific labour. Thus, they gain big margins. In 2004, the USA senate passed another law called Greater Access to Affordable Medicine Act. This act diluted the Hatch-Waxman Act, 1984 in some of the pro-innovator provisions, thereby giving a further boost to the business of generic formulations in the USA.

B. **Increasing outsourcing by MNC pharmaceutical firms:** The contract research and manufacturing services (CRAM) market has offered a great opportunity to the Indian pharmaceutical industry.

India's success in outsourcing to developed countries is due to its low-cost advantage, and world-class quality standards followed by India like internationally harmonized standards of Good Clinical Practices (GCP), current Good Manufacturing Practices (cGMP) & Good Laboratory Practices (GLP). The pharmaceutical industry worldwide is facing challenges of cost pressures for various reasons. In recent years, the R&D productivity of many firms has dropped significantly due to increasing costs of manpower and greater regulatory risk. The approval process for new products in regulated markets needs strict compliance with stringent quality norms, which is also prone to high legal risk. This forces MNCs to outsource some of their manufacturing & R&D activities to destinations like China & India where costs are low. The diversity of disease profile and availability of a large patients base in India makes it an ideal platform for clinical trials. Most of the deals in contract manufacturing are for the manufacturing of active pharmaceutical ingredients (APIs) and intermediates, where India possesses an advantage.

C. **Acquisition by Indian Pharmaceutical firms: acquisition emerged as the main mode of overseas expansion:** The acquisition is the preferred route for international expansion in developed countries as part of the internationalization strategy when compared to organic routes. The benefits of the acquisition are generated by synergies created by the assets supplied by foreign firms & the pipeline of products from an Indian company. Indian companies possess a large product pipeline, low-cost manufacturing plants, and a strong desire to enter the developed countries' markets like the USA & Europe. Major acquisitions occur in the marketing area. Few companies are also investing in developed markets for building capacities in the area of R&D and manufacturing.

Indian firms like Sun Pharmaceuticals Ltd, Wockhardt Ltd & Dr. Reddy's Laboratories have set up manufacturing facilities in

Europe, the USA, Russia, China & Brazil. They have created their brand image in the overseas markets and are further consolidating their activities. To overcome the rising costs of marketing in regulated markets, they are setting up 100% owned subsidiaries or are acquiring promising local firms. This reinforces the internationalization of Indian companies through acquisition, by acquiring knowledge in different areas like regulatory skills, R&D capabilities, and distribution networks.

DRL's acquisition of Betapharm provided them access to the growing European generic drugs market. The product strength of DRL & front-end presence of Betapharm has leveraged DRL's domestic manufacturing advantage.

Indian pharmaceutical companies are acquiring more firms in Europe than in the USA. This is due to price controls of the European government and other regulations which are promoting the use of generic drugs in Europe. Also, in Europe companies are available in a wider price range as compared to the USA where it is more expensive and riskier. Hence, Indian companies have taken the acquisition route for entering the European market for generics but for the generic market in the USA, firms take an organic route.

MARKETING MIX FOR INTERNATIONAL MARKETING

As mentioned earlier, due to uncontrollable P.E.S.T.L.E factors, the international marketer has to adapt controllable elements of the marketing mix to ensure marketing success. We will discuss below how we can make changes to the marketing mix in line with the market environment.

1. **Product & Packaging:** In international markets, decisions related to the quality, packaging, and labeling of products require special consideration. Each country prescribes separate regulations for packaging and labeling, which have to be followed by an international marketing firm. Besides protection and preservation of the goods, packaging and labeling should meet the regulations of the importing countries. Packaging & labeling are often adapted to suit the socio-cultural factors of the target market. Packaging also plays a significant role in the promotion of the product.

   In the case of pharmaceuticals, prior registration of products with the FDA or similar organizations is a must. The quality of the products should meet the regulatory requirements, similarly, there are statutory requirements for product labeling & packaging. E.g., Products with I.P (Indian Pharmacopeia) standards will not be acceptable in overseas markets. For the USA, products must have U.S.P. (U.S. Pharmacopeia) standard while B.P. (British Pharmacopeia) standard will be required for Europe & other Rest of the World. Stock-keeping units will vary from market to market. The Middle East emphasizes the use of Arabic, while Europeans insist on multi-lingual labeling. In Africa, where a lot of self-medication happens, bottles & cartons show pictures to identify the use of medicines. E.g., Picture of mosquito on the anti-malarial drug.

2. **Pricing:** Prices vary across the markets based on competition, affordability of consumers, and government regulations. In regulated markets of USA, Europe & Japan, the prices for Branded & Patented drugs are very high. India mainly supplies 'generics' where prices are relatively higher than in India. However, there is constant pressure from National Health Service & health insurance companies to lower the prices. Also, competition has become intense mainly from Indian suppliers. In the Rest of the World or Semi-regulated markets, it is 3-tiered pricing. For products coming from regulated markets prices are high, those manufactured locally are priced quite low, while Indian medicines are priced in between these two. Again, the completion is intense mainly from Indian suppliers. In some markets like Sri Lanka; there is a 'Price Control' on medicines which limits the profit margin. Also, most of the requirements are bought by Govt. on a 'Global Tender' basis, there is global completion.

3. **Place (Distribution):** Distribution models vary across the regions. In Asia-Pacific & Gulf region, there are set distribution networks like in India. In a country such as the Philippines, there are large 'National Distributors' who service national marketing firms. However, in many countries of Africa, there is a lack of proper distribution network & the marketing firm has to set up distribution warehouses, a fleet of delivery vans & salesmen, develop relations with retailers & manage receivables. In regulated markets of USA & Europe, apart from established distributors, there is a trend for 'Direct supplies' to Hospitals & customers by the marketing firm.

4. **Promotion:** Methods of promotion for pharmaceutical products vary across the regions. In Asia-Pacific & Gulf region, prescription products are typically promoted through the medical profession with the usual modes of communication. OTC products are promoted like consumer items using mass media. In African markets, specialty items like Oncology are promoted through the

medical profession. However, in this market demand is driven by pharmacy/chemists/retailers. Pharmacies stock limited brands & routinely undertake brand substitution of prescribed medicines. Among the patients, there is a large practice of 'Self-medication'. Patients consult chemists & get desired medicines. The packaging of products shows pictures to confirm product usage. E.g., Anti-malarial, Anthelmintic which shows a mosquito or a worm on a pack. In regulated markets like Europe & USA, there is a trend of 'Direct Marketing' wherein marketing firms directly communicate to consumers about the products that are prescribed by the medical profession only. But pre-information of consumers influences the decision-making of prescribers. The standard practice of promotion through a medical representative is followed across the regions, making use of conferences, scientific journals, sampling, and training as promotional tools. With the advent of the internet, there is extensive use of electronic media, websites, e-mail, and SMS; webinars are used to communicate with the medical profession.

PRODUCT REGISTRATIONS

Pharmaceuticals all over the world are heavily regulated products. Drugs & pharmaceuticals are subjected to several regulations and laws concerning the marketing of drugs, ensuring safety & efficacy, patenting, and testing.

There are over 200 countries in the world, and many of them have set up legislation and regulatory requirements for pharmaceuticals. For submission of a regulatory dossier worldwide, it is a mandatory requirement to know guidelines and norms specific to each country. Hence, it is critical to analyze the commonness and differences between the pharmaceutical legislation and regulatory requirements of various countries in the world.

The regulated market countries are those where there are well-defined regulatory requirements established by the regulatory bodies of those countries and the emerging market/semi-regulated countries are the ones who are still lagging in setting up well-defined drugs regulations. EU and United States (US) are the largest and the most potential markets in the world. These are categorized as the regulated markets, while the Rest of the World (ROW) markets cover all the emerging economies like Russia & CIS, Tanzania (Africa), Brazil (LATAM), Hong Kong (ASIA), etc.

A. **Product Registration in the USA:** Regulatory Authority: U.S. Food & Drug Administration.

The Centre for Drug Evaluation and Research (CDER) regulates over-the-counter and prescription drugs, including biological therapeutics and generic drugs.

B. **Product Registration in Europe:** The European Medicines Agency (EMA) is a decentralized agency of the European Union (EU), located in London. The Agency is responsible for the scientific evaluation, supervision, and safety monitoring of medicines in the EU. EMA protects public and animal health in 28 EU Member States, as well as the countries of the European Economic Area, by ensuring that all medicines available on the EU market are safe, effective, and of high quality.

C. **Product Registration in ROW:** Process of Product registration in (ROW) rest of world has become a difficult task just like in regulated markets (US, EU & Japan) because the processes are not harmonized. This area comprises countries from Africa, Latin America, Asia Pacific, Gulf countries, and Eastern Europe. Countries from Gulf and Asia pacific have to some extent harmonized the regulatory environment via the Gulf Co-operation Council (GCC) and The Association of Southeast Asian Nations (ASEAN) organizations. But other ROW nations are yet to harmonize regulations in their respective areas. The immediate need to harmonize & rationalize regulation has arisen due to rising health care costs, R&D, and to satisfy the public requirement for efficacious and safe treatments for needy patients. The commercial importance of the ROW region is growing globally.

In the export registration process, systematic development of formulation becomes a backbone for the preparation of the dossier.

GUIDELINES

1. In the case of various export markets, one can generate a good quality dossier by a systematic development of formulations.

2. Proper process for planning as well as execution of systematic development of formulation will enable preparation of quality dossier & also respond to queries raised by regulatory authorities.

3. As the world is divided based on the requirement of technical data & drug approval procedures, it is vital for generics drugs producers, to carefully understand the target regions, regulatory requirements, market needs, & development costs before the drugs development.

Therefore, it is crucial to plan and coordinate all these activities for timely & successful product launches in the countries.

4. Because of the huge difference in regulatory requirements, it is not practical to seek global marketing approval simultaneously and launch the products at once for all the regions. Given this, we must carefully define & understand the regulatory strategy by considering the different patent terms and their extension, target regions, data requirements, various application possibilities, and possible timelines for launch in different markets. It minimizes the delay in drug approvals, reduces the overall cost of development, and eliminates unnecessary studies and delays in the subsequent launch.

5. The requirements for registration for different countries are available at www.pharmexcil.com in the 'Market & Regulatory Reports' section.

SUMMARY

This chapter deals with international marketing and the international marketing environment, the global pharmaceutical market and role of the Indian pharmaceutical industry, global pharmaceutical market classification, selection of international market, internationalization of the Indian pharmaceutical industry, how the marketing mix for international marketing is different and registration of products in international markets. Key highlights are as follows: -

1. International marketing is defined as the marketing activities carried out across national boundaries. The basic principles of marketing remain the same whether it is domestic or international marketing. However, differences in factors like political, economic, socio-cultural, legal, technological & environment make international marketing a distinct discipline. In the case of the pharmaceutical industry, there is an additional tough requirement to deal with i.e., regulatory affairs & control systems of each country.

2. The international environment in terms of political stability, legal requirements, economic condition, socio-cultural aspects, and technological and environmental factors greatly influence how international marketing is to be carried out.

3. According to IQVIA Market Prognosis, Sep 2021; IQVIA Institute, Nov 2021, the global spending on medicines in 2021 is 1423.5 USD and is estimated to increase by nearly $350 billion, lifting spending to nearly USD 1.8 trillion in 2026.

4. Currently, India is ranked as the 11th largest market and is expected to be ranked as the 9th largest by 2026

5. During the year 2020-21, India's pharma exports were US$24.47 bn. consisting of finished formulations, bulk drugs, surgical items, and Ayush & herbal products.

6. During the year 2020-21 Indian Pharmaceutical industry manufactured generic forms of finished dosages worth US$40.85 bn. India manufactures generic drugs around 20-24 % of the total world's production by value.

7. India has a competitive advantage as a nation in generic formulations due to favorable factors like state-of-the-art manufacturing facilities, a large pool of technically qualified manpower, a large raw material base & low cost of operations, and English language advantage.

8. The Pharmaceutical market based on the diversity in the regulation and marketing interests can be divided into two groups - Regulated markets and Emerging markets (Semi–regulated).

9. Selection of international market depends on market size, accessibility to international markets, profitability, geographical distance, economic distance, political distance, and psychic distance

10. Indian pharmaceutical companies faced domestic competition from global firms and were required to upgrade their capabilities for survival. In 2005, the Indian patent act, of 1970 was amended so that India became compliant with TRIPs (Trade-Related Aspects of Intellectual Property rights), an important for pharmaceutical products in India. Many Indian pharmaceutical companies have further moved up in the value chain & thus internationalized by acquiring small firms and also setting up wholly-owned subsidiaries, to access resources and enter newer markets.

11. Developments that favored internationalization of the Indian pharmaceutical industry were the Hatch-Waxman Act & opportunities for generic drugs in the USA, increasing outsourcing by MNC pharmaceutical firms, and acquisition by Indian pharmaceutical firms.

12. The product/packaging, pricing, distribution, and promotion in international marketing varies as per the country's regulations, local trends, and marketing practices in those countries.

13. Pharmaceuticals all over the world are heavily regulated products. Drugs & pharmaceuticals are subjected to several regulations and laws concerning the marketing of drugs, ensuring safety & efficacy, patenting, and testing.

14. For submission of a regulatory dossier worldwide, it is a mandatory requirement to know guidelines and norms specific to each country.

REFERENCES

1. Joshi Rakesh Mohan, International Marketing, 2nd Edition, Oxford University Press, 2014.

2. 17th Annual Report 2020-21, Pharmaceuticals Export Promotion of India, Ministry of Commerce & Industry, Govt. of India.

3. "Pharmaceuticals Global Market Report 2021: COVID-19 Impact and Recovery to 2030", Research and Markets, Dublin, March 31, 2021 (GLOBE NEWSWIRE)

4. Jitendra Kumar Badjatya, Ramesh Bodla, Pankaj Musyuni, Export Registration of Pharmaceuticals in Rest of World countries, Journal of Drug Delivery & Therapeutics; 2013, 3(1), 61-64.

5. Kale Dinar, Internationalisation strategies of Indian pharmaceutical firms. *PharmaBuzz*, 2010, 5(10) pp. 6–13.

Chapter **22**

OTC MARKETING – A GROWTH STRATEGY

LEARNING OUTCOMES

After reading this chapter, you should be able to:

- Understand the concept of OTC Marketing

- Understand the significance of OTC Marketing

- Get an insight into the challenges of OTC Marketing in India

- Understand the process of OTC Marketing

- Formulate Rx to OTC strategies

OTC MARKETING

With the increasing health consciousness, more and more people in India are adopting self-medication. Self-medication is the treatment of common health problems with medicines especially designed and labeled for use without medical supervision and approved as safe and effective for such use.

Medicines for self-medication are often called 'nonprescription' or over-the-counter (OTC) and are available without a doctor's prescription (Rx) through pharmacies. 'OTC Drugs' in common parlance means drugs that are legally allowed to be sold Over the Counter without the prescription of a Registered Medical Practitioner. In the Indian context, the phrase 'Over the Counter' and the abbreviation 'OTC' are also referred to as nonprescription drugs. There is a need to build a robust regulatory framework for OTC medicines in India and the government is working towards this goal. India will soon allow the sale of certain drugs without the requirement of a prescription, as part of a new over-the-counter (OTC) policy for drugs. The Drugs Technical Advisory Board (DTAB), has approved a list of OTC drugs which includes medicines like antifungal infections, analgesics (pain killers), cough syrups, decongestants, laxatives, antiseptics, and medicines for gum infections (The Economic Times, Jan 20, 2022).

SIGNIFICANCE OF OTC MARKETING IN INDIA

Indian OTC market has been valued at approx. Rs.30, 000 crores with a CAGR (compounded annual growth rate) of 9 percent (Dec'20 MAT, as per Nicholas Hall's DB6 Global OTC Database – India).

Self-medication is on the rise, especially when it comes to minor ailments like headaches, cough, cold, digestive problems, etc. as revealed by Ipsos pan-India survey (ETHealthworld.com, June 03, 2021). Increased literacy, paucity of time, high doctor fees, mass media promotion of OTC products by pharmaceutical companies, and the emergence of new categories like cosmeceuticals, health supplements, and wearable devices are some reasons for the rise in self-medication. There is tough competition among prescription products as multiple brands of similar composition are available and each brand is vying for the share of doctors' prescriptions. Shifting the brand from prescription to OTC allows the pharmaceutical company to reduce its dependency on doctors and also to further build the brand.

In the case of patented products, the pharmaceutical company enjoys the monopoly till the patent period is valid. However, once the patent period expires, it leads to a rise in generic competition resulting in a decline in market share and profits. Thus, by proactively shifting the product from prescription to OTC, the pharmaceutical company can extend its product life cycle.

CHALLENGES OF OTC MARKETING IN INDIA

Although there is a huge opportunity for OTC marketing in India, there are many challenges as well. These challenges are as follows:-

1. Distribution
2. Product formulation
3. Advertising

4. Field Force

5. Consumer Psychology

6. Pharma mindset

1. **Distribution:** The exploitation of the OTC opportunity by a new entrant will depend on setting up an effective channel of distribution network. Apart from widening their product profile and offerings, companies need to increase the market reach of their products. It is a general prophecy that the advent of organized retail will create new shelf space. Pharmaceutical companies need to go beyond the 'chemists and druggists' to make their products easily accessible to the consumers. Out of 12 million retail outlets in India (the total retail universe), only 6-7 percent are drug stores. So, there is a need to reach out to retail outlets which are non-pharmacy outlets. To increase sales of an OTC product its distribution has to be increased for which in addition to pharmaceutical distributors even FMCG distributors should be brought in since their reach is far wider. Thus for the pharmaceutical companies which are used to distribution only through chemists, extending beyond the chemists' outlets becomes a challenge.

2. **Product formulation:** An OTC product needs to be formulated in such a manner that it is out of schedule. Most companies in India are trying to overcome this by adopting an herbal route.

   Secondly, an OTC product needs to be consumer-friendly so that it is easier for consumers to buy on their own and consume it. This calls for extensive consumer research to understand consumer needs and pain points.

   Both the above points pose a challenge for the pharmaceutical company.

3. **Advertising:** Consumption of an OTC brand is dependent on brand awareness, which in turn is dependent on media advertising (Vijay Bhangale, 2013). Awareness through media advertising is one of the most crucial & important steps in OTC Marketing. The OTC products need to be mass advertised like the FMCG products.

   Prescriptions products promoted through scientific detailing to doctors are not advertised in mass media, so the Indian pharmaceutical companies are not much aware of the media & advertising concepts & procedures. Thus, developing consumer communication and advertising in mass media becomes a challenge.

4. **Field force:** This is one of the biggest challenges for Indian pharmaceutical companies. The pharmaceutical field force is

inclined towards promoting products to doctors only whereas for OTC marketing the field force needs to promote products to consumers and generate business through retailers. There is a need to change the mindset from a doctor detailing to retailing.

A new set of field forces needs to be created for OTC marketing to service the distribution channel and the retail outlets

5. **Consumers:** Several challenges for OTC Marketing in India from the consumer point of view are as follows-

Low consumer awareness and knowledge about ailments and medicines

Consumer inertia - consumers think that it's a minor problem and will go on its own

There is a trust deficit among consumers concerning OTC medicines

No clarity about usage occasion, dosage, etc.

6. **Pharma mindset:** The typical pharma mindset is Low investment & Quick returns as the promotion is restricted to healthcare professionals only. OTC marketing on the other hand calls for huge investment which is for the long term as the promotion is direct to consumers and in mass media.

PROCESS OF OTC MARKETING IN INDIA

There is no defined process for Rx to OTC switch in India. Given below are important criteria that need to be taken into account before switching a prescription brand to OTC.

CRITERIA FOR THE "RX TO OTC" SWITCH

(Adapted from the book Brand Medicine, Tom Blackett and Rebecca Robins, 2001)

1. **Regulation:** The first & foremost important criteria would be to see that the brand which the company is keen to aggressively promote OTC is not in the scheduled category (schedule H, X, G) and hence not governed by any government regulations.

2. **Self-medication potential:** Secondly, analyze and evaluate whether the brand has the potential to go DTC (Direct-To-consumer) i.e. does it have the 'self-medication' kind of profile. E.g. "Analgesics & Antacids" do have great self-medication potential.

3. **Current market size:** It is important to evaluate the current market size of the therapeutic category the brand belongs to. The current market size should be large and provide enough growth opportunities.

4. **Future growth potential:** Need to check what future growth potential the brand has. The growth prospects in the prescription market may be less (for e.g doctors may switch over to newer better drugs) but may have good potential to grow if promoted OTC as there is a growing trend among consumers to consume the brand on their own.

5. **Price regulation:** Pricing of the brand should not be governed by DPCO (Drug Price control Order). The reason is price control limits the profit margins and thereby the investment in Advertising & Sales Promotion.

6. **Competition:** Need to have a holistic view of the competition. The competition will not be restricted to the therapeutic category but could also be from alternative medicines, home remedies, etc.

7. **Distribution perspective:** The distribution orientation needs to be changed to FMCG type of distribution i.e. the brand should be readily available in pharmacy as well as non-pharmacy outlets.

8. **Organizational capability:** The organization should have enough resources to invest in mass advertising & promotions. Also, the skills & competencies in Marketing Research, Brand Development, and Sales & Distribution would be necessary.

9. **Technical & commercial capabilities:** Need to check on the technical & commercial aspects like:

    Technical: Should be able to develop consumer-friendly formulations e.g. innovative products or delivery systems to attract the consumers.

    Commercial: Cost efficiencies, profitability, etc.

10. **Brand life cycle:** Consider the life cycle stage of the brand. E.g. once the drug goes off-patent there is competition from generics & prescription market becomes overcrowded resulting in declining market share & profits. 'OTC Promotion' may be one of the strategies to build the brand further.

11. **Mind-share of the brand:** Need to assess the mind share of the brand amongst doctors & consumers.

12. **Brand heritage:** It is important to have an idea about consumers' perception of the brand. Do they think it's young and vibrant or old, outdated?

13. **Profitability:** Need to assess the profitability. Rx to OTC switch involves a very high investment & also a high risk. During the switch, it needs to sustain the bottom line.

14. **Rx-base vs repeat Purchase:** One of the most important criteria is whether brand sales are due to heavy prescription or repeat purchases. If the sales contribution is more due to the repeat purchase then the switch may be a wise decision as the consumers are already aware as they are buying on their own.

Rx to OTC Switch Strategies

Let's understand some of the key Rx to OTC switch strategies with the help of an example of Rx to OTC switch of the brand, Revital (a health supplement that is a combination of vitamins, minerals, and ginseng). The strategies are as follows:-

1. **Communicate real consumer benefit – a shift from serious medical image:** Since the target consumer for an OTC brand is a layman, it is important to communicate the benefit that the consumer will easily understand and relate to. When Revital shifted from Rx to OTC it targeted the consumers leading a hectic lifestyle, empowering them with energy strength, and mental sharpness. Extensive consumer research was carried out to gather consumer insights and accordingly, an advertising campaign was developed highlighting that Revital enables consumers to live life to the fullest by reducing fatigue and weakness.

2. **Brand image & Brand building:** Revital product packaging was made consumer-friendly and the brand-building exercise was carried out to increase awareness, and build image and loyalty. Besides advertising in mass media, Revital carried out sampling and various consumer activation programs at different consumer touchpoints.

3. **Value of a well-known name:** Revital was prescribed by doctors to several patients and also there were repeat purchases by consumers on their own. Since the brand enjoyed good medical equity and also had fairly good awareness among the consumers, it was beneficial to continue with the same brand name.

4. **Advertising:** Since OTC product purchase is similar to a fast-moving consumer good purchase, advertising in mass media is

important to increase awareness, build image and enhance recall. Revital is extensively advertised on mass media like television, newspaper, magazines, out-of-home along with online media.

5. **Extensive distribution and the pivotal role of retailer:** OTC products need to follow the FMCG type of distribution network as it has to be readily available and therefore should be available in chemist outlets as well as other outlets like general stores, departmental stores, modern retail, etc. Secondly, it is important to educate the retailer about the product benefits, the condition in which it is to be used, dosage, precautions, etc. Revital was distributed by pharma as well as FMCG distributors and was made available in chemist outlets as well as general stores, departmental stores, modern retail, etc. An extensive retailer education drive was carried out and consumer education leaflets were also kept at retail counters to be handed over to the consumers.

6. **Continue promotion to doctors:** Endorsement by doctors helps to build trust amongst consumers. Hence it is important to retain the medical equity of the Rx brand even after its switch to OTC.

DEVELOPING AN ADVERTISING CAMPAIGN

In developing an advertising campaign, marketing managers must always start by identifying the target market and buyer motives.

Different stages are as follows: -

1. **Planning & information review**
   - Specify the need for advertising, budget available, and time available
   - Gather product insights and consumer insights

2. **Develop creative brief**
   - What is the Ad requirement
     - What is the purpose of this brief? (launch)
     - How many executions are required, and for what medium?
   - What problem Ad must solve
     - What key aspects of consumer behavior or beliefs must be addressed (fix lack of awareness; tackle below avg penetration amongst the particular target group.)

- Creative Strategy
  - Market Description, Target audience, Key Consumer Insight, Point of Differentiation, Positioning, Reason to Believe, Advertising Idea, Brand Personality
- Required Consumer Response
  - How we want consumers to change (gain trial, remain loyal, use different occasions)
- Mandatory Executionary Requirements
  - Visuals, Legal/Regulatory constraints

3. **Hold creative presentation meeting, with senior people present**
    - Evaluation on
      - Strategy (selling idea, how you expect consumers to respond to Ad, is Ad on strategy)
      - Overall impression (like/dislike, persuasion, impact)
      - Is the Selling idea clear & central (will consumers understand it the first time, is it central to the commercial or just a wrap-up line)
      - Is the benefit visualized (ask whether pictures alone communicate the selling idea)?
      - Is Ad focused on Brand/Product (good branding, packshot)
      - Does it have dramatic interest (is the product 'hero' of the Ad)
      - Is it simple?
      - Does it have relevance
    - Confirm whether creative submission affects the media plan or vice versa
    - Check with Legal, Regulatory
    - Confirm Pre-testing plan
    - For Pre-testing, develop an animatic of the board
    - Review Pre-test results
    - Select Ad for production

4. **Manage production process & final testing**
    - Get approvals from management, legal
    - For TV, hold a "Storyboard meeting"

- Creative Agency selects Director and gets production house quotes
- "Pre-Production" meeting
- Shoot
- Agree on a rough edit
- Agree finished Ad
- Research Finished Ad

5. **Start advertising & measure its effectiveness**
   1. Confirm Start date & creative rotation (if applicable)
   2. Confirm how to measure the effectiveness of new execution

MEDIA SELECTION & PLANNING

Media selection involves finding the most cost-effective media to deliver the desired number of exposures to the target audience.

The next task is to find out how many exposures, will produce a level of audience awareness. The effect of exposures on audience awareness depends on the exposures' reach, frequency, and impact.

Reach The number of different persons or households exposed to a particular media schedule at least once during a specified period.

Frequency: The number of times within the specified period that an average person or household is exposed to the message.

Impact: The qualitative value of an exposure through a given medium

The media planner has to figure out, with a given budget, the most cost-effective combination of reach, frequency, and impact.

Advertisers need to make decisions at each of three successive levels in selecting the specific advertising medium to use:

1. Which *type(s)* will be used-newspaper, television, radio, magazine, internet, or direct mail?
2. Which *category of the selected medium* will be used? Television has network and cable; magazines include general-interest (India Today) and special-interest categories; there are national as well as local newspapers, and the Internet offers portals as well as individual websites.

3. Which *specific media vehicles* will be used?

The major media vehicles available are Television, Newspapers, Radio, Magazines, Direct Mail, Brochures, the Internet, Mobile, and Out - Of - Home (Billboards, Buses, Metro, etc.). All of these media vehicles have their respective pros & cons in terms of coverage, effectiveness, cost, lead time required, target audience selectivity, etc.

Here are some general factors that will influence media choice:

- *The objective of the Ad: -P*urpose of a particular ad and the goals of the entire campaign influence which media to use. For example, if the campaign goal is to generate appointments for salespeople, the company may rely on direct mail. If an advertiser has a short lead time, a local newspaper or radio may be the medium to use.

- *Audience Coverage: -.* The audience reached by the medium should match the geographic area in which the product is distributed. Furthermore, the selected medium should reach the desired types of prospects with a minimum of wasted coverage that reach people who are not prospects for the product. Many media, even national and other large-market media can be targeted at small, specialized market segments.

- *Requirements of the message: -*The medium should fit the message. For example, magazines provide high-quality visual reproductions that attract attention along with printed messages that can be carefully read and evaluated; As a result, they are well suited to business-to-business advertising.

- *Time and location of the buying decision*: - If the objective is to stimulate a purchase, the medium should reach prospective customers when and where they are about to make their buying decisions. This factor highlights one of the strengths of point-of-purchase advertising (such as ads placed on shopping carts and in the ~ aisles of supermarkets), which reaches consumers at the actual time of purchase.

- *Media cost*: - The cost of each medium should be considered taking into account the funds available and its reach or circulation. For example, the cost of network television exceeds the available funds of many advertisers. To compare various media, advertisers use a measure called cost per thousand (CPM), which is the cost of reaching a thousand people, one time each, with a particular ad.

Beyond these general factors, management must evaluate the advertising characteristic of each medium it is considering. The term characteristic is chosen, instead of advantages and disadvantages, because a medium that works well for one product is not necessarily the best choice for another product. To illustrate, a characteristic of radio is that it makes its impressions through sound and imagination.

SUMMARY

This chapter deals with an understanding of the concept of OTC Marketing, its significance and challenges in the Indian context, the process of OTC Marketing, and Rx to OTC switch strategies. Key highlights are as follows:-

1. 'OTC Drugs' in common parlance means drugs that are legally allowed to be sold Over the Counter without the prescription of a Registered Medical Practitioner. In the Indian context, the phrase 'Over the Counter' and the abbreviation 'OTC' are also referred to as nonprescription drugs. There is a need to build a robust regulatory framework for OTC medicines in India and the government is working towards this goal.

2. Indian OTC market has been valued at approx. Rs.30, 000 crores with a CAGR (compounded annual growth rate) of 9 percent (Dec'20 MAT, as per Nicholas Hall's DB6 Global OTC Database – India).

3. Self-medication is on the rise, especially when it comes to minor ailments like headaches, cough, cold, digestive problems, etc. Increased literacy, paucity of time, high doctor fees, mass media promotion of OTC products by pharmaceutical companies, and the emergence of new categories like cosmeceuticals, health supplements, and wearable devices are some reasons for the rise in self-medication

4. There is tough competition among prescription products as multiple brands of similar composition are available and each brand is vying for the share of doctors' prescriptions. Shifting the brand from prescription to OTC allows the pharmaceutical company to reduce its dependency on doctors and also to further build the brand.

5. Setting up FMCG type of distribution, formulating consumer-friendly products, consumer advertising, field force promoting products to retailers and consumers, consumer psychology concerning OTC medicines, and pharma company mindset are the key challenges for OTC marketing in India.

6. While switching from Rx to OTC one needs to check on these criteria - regulation, self-medication potential, current market size, future growth potential, price regulation, competition, distribution perspective, organizational capability, technical & commercial

capabilities, brand life cycle, mind-share of the brand, brand heritage, profitability and Rx base vs repeat purchase.

7. Some of the key Rx to OTC switch strategies are communicating the real consumer benefit – a shift from a serious medical image, brand image & brand building, value of a well-known name, advertising in mass media, extensive distribution, and the pivotal role of the retailer. Continue doctor promotion as retaining medical equity is important to build consumer trust.

8. The process of developing an advertising campaign includes planning & information review, developing a creative brief, holding a creative presentation meeting, managing the production process & final testing, starting media advertising & measuring its effectiveness.

9. Media selection and planning involve finding the most cost-effective media to deliver the desired number of exposures to the target audience. The next task is to find out how many exposures, will produce a level of audience awareness. The effect of exposures on audience awareness depends on the exposures' reach, frequency, and impact.

REFERENCES

1. Thacker Teena, Govt to allow the sale of some drugs sans prescription, The Economic Times, Jan 20, 2022, available at https://economictimes.indiatimes.com/news/india/govt-to-allow-sale-of-some-drugs-sans-prescription/articleshow/89007321.cms?utm_source=contentofinterest&utm_medium=text&utm_campaign=cppst

2. Nicholas Hall's DB6 Global OTC Database – India, Dec'20 MAT

3. ETHealthworld.com, June 03, 2021, https://health.economictimes.indiatimes.com/news/industry/headache-most-common-health-ailment-among-urban-indians-survey/83200100 accessed on Sept 5, 2021.

4. Bhangale Vijay, A study on the impact of media advertising on the consumption of OTC cough and cold medicines, International Journal of Healthcare Management, 2013, Vol. 6 No. 2, pg 136-140.

5. Blackett Tom and Robins Rebecca, Brand Medicine, The Role of Branding in the Pharmaceutical Industry, 2001, Palgrave, New York.

MARKETING OF SPECIALTY DRUGS FOR RARE DISEASES

LEARNING OUTCOMES

After reading this chapter, you should be able to:

- Understand Specialty drugs

- Understand Rare diseases

- Marketing of specialty drugs for rare diseases

SPECIALTY DRUGS

A specialty drug is defined as any drug that requires "high-touch" service in the form of distribution, administration, or patient management — all factors that drive up costs for the purchaser or consumer (Scott Kober, Biotechnology Healthcare, Jul-Aug 2008). Specialty drugs are designed to treat rare, complex diseases. From the 1990s to the 2000s, the number of specialty drugs has grown from 30 to over 500 (www.truveris.com). By 2023, 65% of new drug launches are expected to be specialty drugs (IQVIA report on Global use of medicine in 2019 and outlook to 2023, Jan 29, 2019).

Characteristics of specialty drugs are as follows (adapted from www.truveris.com):-

1. Specialty medications are typically biologic medications, meaning the drug is made from living organisms and targets a specific part of the disease process.

2. Specialty drugs treat certain cancers, multiple sclerosis, narcolepsy, and other care-intensive conditions.

3. In contrast to specialty medications, non-specialty drugs are typically small-molecule medications, meaning they are chemically synthesized. The most common non-specialty medications treat conditions like cardiovascular disease, high blood pressure, and diabetes. The non-specialty drugs treat both chronic and acute diseases that affect larger populations.

4. While specialty drugs were originally designed to treat only rare diseases, as time has progressed, specialty drugs are increasingly being developed to treat common chronic diseases, such as asthma and eczema.

5. Specialty drugs require intensive packaging, handling, storing, and monitoring, which is why their distribution is restricted. For example, some specialty drugs must be kept at a certain temperature, others require significant monitoring and data collection, usually mandated by the regulatory authority which further limits the number of pharmacies that are permitted to dispense a drug.

6. Specialty drugs are very expensive as compared to non-specialty drugs.

RARE DISEASES

Rare diseases, as the name suggests are the diseases that very few people suffer from within a population. To address various challenges in the management and research of rare diseases, a comprehensive "National Policy for Rare Diseases 2021" was approved by the Ministry of Health and Family Welfare, Government of India in March 2021. World Health Organization (WHO) defines a rare disease as often debilitating lifelong disease or disorder with a prevalence of 1 or less, per 1000 population (National policy for rare diseases, 2021). Drugs Controller General of India (DCGI), in March 2019, came out with a definition of orphan drugs, which essentially are used to treat rare diseases, as the one which addresses a potentially treatable population of 500,000 in India. Rare diseases include inherited cancers, autoimmune disorders, congenital malformations, and certain endemic infectious diseases due to their rarity and low prevalence (Orphan et, n.d.). It is estimated that there are 7,000 rare diseases, of which 80% are genetic and only 5% are treatable (National Institutes of Health, n.d.). A single rare disease affects a small number of people, but rare diseases collectively affect millions. Over 70 million people in India suffer from rare diseases. About 700 rare or ultrarare diseases are currently present in India (Rare Diseases India, n.d.).

MARKETING OF SPECIALTY DRUGS FOR RARE DISEASES

Marketing specialty drugs for rare diseases is unique as the biggest challenge is in identifying and diagnosing patients. Families who experience rare diseases need more than product information; they need education and support. They need tools to teach themselves, and others, about daily disease management, while also advocating for effective treatment (PharmExec.com, January 31, 2020).

There are four key factors that life-sciences companies need to keep in mind when developing relationships with the specialty physician (PharmaVoice, Oct 2016) as follows: -

1. Increased integration around specific therapeutic areas to allow a more integrated engagement model around the patient with the help of technology.

2. Companies should start shifting investments away from traditional marketing initiatives to value-based selling, e.g., Health Economics and Outcomes Research (HEOR) data generation down to specific key accounts.

3. To capture enterprise-level customer insights to drive better coordination across medical, market access, and marketing and sales teams. This can only work if cloud-based technologies are used to integrate a variety of data sets continuously and in real-time.

4. Companies should consider new research that shows the tremendous impact of localization of marketing and sales strategies, tailored to local market dynamics.

Rare diseases are not only difficult to diagnose but are also complicated to treat. Companies that launch therapeutics in this area are bold and their branding should reflect that (PharmExec.com, July 1, 2021). Three ways in which this can be achieved are as follows-

1. **Be an early leader**

   Define and refine your category from the start, it will pay off tenfold during the life cycle of your product.

2. **Engage advocates**

   Partner with those impacted by the disease, including patients, care partners, advocacy organizations, and healthcare practitioners. Find them, talk with them, listen to them, learn from them, and work alongside them.

3. **Own every aspect of the therapeutic category**

   Be the number one expert in the field. Every aspect of the brand, from its name to the way it is presented to healthcare practitioners, should reflect that boldness.

SUMMARY

This chapter deals with the understanding of specialty drugs and rare diseases & marketing of specialty drugs for rare diseases. Key highlights are as follows: -

1. A specialty drug is defined as any drug that requires "high-touch" service in the form of distribution, administration, or patient management — all factors that drive up costs for the purchaser or consumer.

2. Specialty drugs are designed to treat rare, complex diseases.

3. Drugs Controller General of India (DCGI), in March 2019, came out with a definition of orphan drugs, which essentially are used to treat rare diseases, as the one which addresses a potentially treatable population of 500,000 in India. Rare diseases include inherited cancers, autoimmune disorders, congenital malformations, and certain endemic infectious diseases due to their rarity and low prevalence.

4. It is estimated that there are 7,000 rare diseases, of which 80% are genetic and only 5% are treatable. About 700 rare or ultrarare diseases are currently present in India and over 70 million people in India suffer from rare diseases.

5. Marketing specialty drugs for rare diseases is unique as the biggest challenge is in identifying and diagnosing patients. Families who experience rare diseases need more than product information; they need education and support. They need tools to teach themselves, and others, about daily disease management while also advocating for effective treatment.

REFERENCES

1. Kober Scott, The Evolution of Specialty Pharmacy, Biotechnology Healthcare, Jul-Aug 2008, 5(2): 50–51.

2. https://truveris.com/difference-between-specialty-drug/ accessed on Jan 31, 2022.

3. https://www.iqvia.com/insights/the-iqvia-institute/reports/the-global-use-of-medicine-in-2019-and-outlook-to-2023 accessed on Jan 31, 2022.

4. https://main.mohfw.gov.in/sites/default/files/Final%20NPRD%2C%202021.pdf

5. Orphanet. (n.d.) The portal for rare diseases and orphan drugs. https://www.orpha.net/consor/cgi-bin/index.php

6. National Institutes of Health (n.d.) Rare diseases. https://www.nih.gov/about-nih/what-we-do/nih-turning-discovery-into-health/rare-diseases

7. Rare Diseases India. (n.d.). Estimated rare diseases population in South Asian countries. http://www.rarediseasesindia.org/

8. Wilson Laura and Romeo Jonathan, How Rare Disease Marketing Will Stretch Beyond the Patient, PharmExec.com, January 31, 20202.

9. Robinson Robin, The Specialty Drug Market, PharmaVoice, Oct 2016.

10. Cashion Brannon, Budd Vince, 3 Bold Moves to Stand Out in Rare Disease Branding, PharmExec.com, July 1, 2021.

BIOPHARMACEUTICALS

LEARNING OUTCOMES

After reading this chapter, you should be able to:

- Get an insight into Biopharmaceuticals market analysis
- Understand Future scope & Strategies

BIOPHARMACEUTICALS & BIOSIMILARS

The term "biopharmaceuticals" refers to pharmaceuticals produced in biotechnological processes using molecular biology methods (Malgorzata Kesik-Brodacka, 2017). Biopharmaceuticals target only specific molecules, rarely causing the side effects associated with conventional small-molecule drugs (D.J.Craik et.al. 2013). Secondly, they exhibit specificity and activity as compared to conventional drugs (S.Mitragotri et.al. 2014).

Generics are defined as drugs that are equivalents of the innovative reference drugs containing the same active pharmaceutical ingredient. The term refers to substitutes for synthetic drugs. The term "generic" is not used in reference to biopharmaceuticals (Malgorzata Kesik-Brodacka, 2017). The European Medicines Agency (EMA) decided that the term "biosimilars" should be used in the European Union (EU) to refer to biological medicinal products containing a version of the active pharmaceutical ingredient found in previously registered reference biological medicinal products The U.S. Food and Drug Administration (FDA) uses the term follow-on biologics. Moreover, both of these agencies determined that biosimilars may have different actions than the reference drug.

A biosimilar is a biologic that has similar structural and pharmacokinetic properties to the innovator biologic and is capable of providing the same therapeutic effect. Biosimilars can be considered follow-on or generic versions of an innovator biologic.

Challenges with biosimilar biotherapeutic products as compared to generic drugs (Annu Uppal et. al., 26 Nov 2021) are as follows: -

- Biotherapeutics, particularly monoclonal antibodies, are challenging to manufacture, are heterogeneous products, and require a high level of process control for a manufacturer to ensure consistent product quality

- Treatment costs are high with biosimilars. Unlike generic drugs, whose price tags typically reflect a 90% reduction from their branded counterparts, a biosimilar is generally sold at a modest 20–30% reduction in cost relative to the innovator drug

BIOPHARMACEUTICALS MARKET

According to the Iqvia report on 'Spotlight on Biosimilars', June 2021, the global biologic medicines market is valued at $320 billion in 2020 and accounts for almost one-third of the global market for pharmaceuticals by value. In Europe, biologic molecules account for 40% of the total market value in 2020, and new launches are expected to increase from 13 new molecules per year (2014–2018) to 27 per year (2021–2025). By 2025, $112 billion of biologic medicines will lose exclusivity in the global market.

Currently, the market for biologics is very condensed in terms of the number of indications in which they can be used, and is dominated by indications such as autoimmune diseases, oncology, and diabetes, which have a high incidence and prevalence rate. Among these, oncology was valued at USD 136bn in 2018 and is estimated to be valued at USD 220bn by 2024, growing at a CAGR of 8.4% (www.mordorintelligence.com).

The Indian biopharmaceutical industry, in the recent past, delved into biosimilar production. A biosimilar is a biotherapeutic product, similar in terms of quality, safety, and efficacy to an already licensed reference biotherapeutic product. Oversimplifying — they could be considered biologically produced (not chemically synthesized) generics. Biosimilars can alleviate the social and economic differences prevalent in medical care today. These can challenge the reference products by improving patient access through affordability. The market for biosimilars has also been exponentially on the rise in India projected to be $40 billion by 2030 (IndiaBioscience, 2019).

Association of Biotechnology Led Enterprises (ABLE) in India estimates the biologics, biosimilars, and vaccines market to grow at a compounded annual growth rate (CAGR) of 22% to become USD 12bn by 2025(Department of Biotechnology, Government of India, Nov 2019). India is predicted to be one of the world's 'fastest-growing bio-hub' in recent years. India's growing biosimilars industry is the primary driver of growth, with over 98 approved biosimilars in the domestic market. Leading Indian companies are Reliance Lifesciences, Biocon, USV, Serum Institute of India, Zydus Cadila, Cipla, Intas, Lupin, Dr.Reddy's Labs, etc.

STRATEGIES TO BE SUCCESSFUL IN THE BIOPHARMACEUTICAL MARKET

With therapeutic innovation, the use of predictive analytics and artificial intelligence (AI), the increasing role of patients and caregivers, market access restrictions, pricing pressure, and new manufacturing and distribution channels and players, the biopharmaceutical ecosystem is changing. To succeed in this changing environment, biopharmaceutical companies need to adopt innovative strategies. Some of these strategies (adapted from Pharmaceutical Executive, June 2021 & Pwc report on 'Vision to Decision Pharma 2020) are as follows: -

1. **Create more value:** This can be done in the following manner –

   - By engaging with payers and other stakeholders well in advance before the launch to educate them on the disease and unmet needs as well as understand what value means to each stakeholder.

   - Collecting real-world evidence of a medicine's effectiveness

   - Measuring how patients feel and

   - Developing companion diagnostics for specialist therapies.

2. **Meaningful engagement with the expanded set of influencers and decision-makers:** To be successful, biopharma companies need to meaningfully engage patients/caregivers and advocacy groups to understand their needs at a fundamental level and communicate information in highly tailored ways. Since the information being communicated is increasingly complex medical affairs can play a leading role in KOL management. Digital health

technologies and gamification can be made use of for patient education and ensuring patient compliance.

3. **Segmentation based on data analytics:** Artificial intelligence (AI)-based approaches can be applied throughout the entire value chain from drug discovery to patient-finding to commercial marketing to characterize customer archetypes and their unique needs and tailor content and engagement accordingly.

4. **Multi-faceted customer-facing teams:** There is a need for engaging customers through a team of personnel with specialized capabilities. HCPs and their staff are being engaged not only by sales representatives and medical science liaisons (MSLs), but also by commercial thought leader liaisons, field reimbursement specialists, and clinical educators. For payers, many companies now employ medical outcomes specialists as well as traditional commercial account managers, and patients and their caregivers may interact with patient educators, case managers, reimbursement support specialists, and patient advocacy engagement personnel. These field teams need to be highly coordinated to maximize the impact of their collective efforts.

SUMMARY

This chapter deals with the understanding of biopharmaceuticals and biosimilars, market analysis, and strategies to be successful in this market. The key points discussed in this chapter are as follows:

1. The term "biopharmaceuticals" refers to pharmaceuticals produced in biotechnological processes using molecular biology methods.

2. A biosimilar is a biologic that has similar structural and pharmacokinetic properties to the innovator biologic and is capable of providing the same therapeutic effect. Biosimilars can be considered follow-on or generic versions of an innovator biologic.

3. According to the Iqvia report on 'Spotlight on Biosimilars', June 2021, the global biologic medicines market is valued at $320 billion in 2020 and accounts for almost one-third of the global market for pharmaceuticals by value.

4. The market for biosimilars has also been exponentially on the rise in India projected to be $40 billion by 2030 (IndiaBioscience, 2019).

5. Association of Biotechnology Led Enterprises (ABLE) in India estimates the biologics, biosimilars, and vaccines market to grow at a compounded annual growth rate (CAGR) of 22% to become USD

12bn by 2025(Department of Biotechnology, Government of India, Nov 2019). India is predicted to be one of the world's 'fastest-growing bio-hub' in recent years.

6.   Strategies to be successful in the biopharmaceutical market are - to create more value, meaningful engagement with all stakeholders, segmentation based on data analytics, and have multi-faceted customer-facing teams.

REFERENCES

1.   Kesik-Brodacka Malgorzata, Progress in biopharmaceutical development, Biotechnology and Applied Biochemistry, 2 Nov 2017.

2.   Craik, D. J., Fairlie, D. P., Liras, S., and Price, D. (2013) Chem. Biol. Drug Des.81, 136–147.

3.   Mitragotri, S., Burke, P. A., and Langer, R. (2014) Nat. Rev. Drug Discov. 13, 655–672.

4.   Guideline on similar biological medicinal products. Available at http://www.ema.europa.eu/docs/en_GB/document_library/Scientific_guideline/2014/10/WC500176768.pdf. accessed on 2 Feb 2022.

5.   Scientific Considerations in Demonstrating Biosimilarity to a Reference Product. U.S. Department of Health and Human Services Food and Drug Administration. Available at https://www.fda.gov/downloads/Drugs/GuidanceComplianceRegulatoryInformation/Guidances/UCM291128.pdf. Accessed 2 Feb 2022.

6.   Uppal Annu, Chakrabarti Ranjan, Chirmule Narendra, Rathore Anurag, Atouf Fouad, Biopharmaceutical Industry Capability Building in India: Report from a Symposium, Journal of Pharmaceutical Innovation, https://doi.org/10.1007/s12247-021-09596-9

7.   Iqvia report on Spotlight on Biosimilars, June 2021

8.   https://www.mordorintelligence.com/industry-reports/cancer-therapy-market

9.   Kapur Ibani, Jain Navodita, The growing Indian biopharmaceutical industry: its requirements and endeavors, Feb 15, 2019 indiabioscience.org/columns/education/the-growing-Indian-biopharmaceutical-industry-its-requirements-and-endeavors accessed on Jan 2, 2022.

10. Special report on India: The emerging hub for biologics and biosimilars, Department of Biotechnology, Government of India, Nov 2019.

11. Lombardi Bart, Product Launch Strategies: Adapting Approaches in 2020 and Beyond, Pharmaceutical Executive-06-01-2020, Volume 40, Issue 6.